The BUILDING WORK HANDBOOK

A Practical Guide for Contractors and Clients

Robert W. Howe

GUILD OF
MASTER CRAFTSMAN
PUBLICATIONS

First published 2010 by
Guild of Master Craftsman Publications Ltd.
Castle Place, 166 High Street, Lewes,
East Sussex BN7 1XU

ISBN 978 1 86108 689 1

A catalogue record for this book is available from the British Library.

Associate Publisher Jonathan Bailey
Production Manager Jim Bulley
Managing Editor Gerrie Purcell
Senior Project Editor Dominique Page
Editor Guy Croton
Managing Art Editor Gilda Pacitti
Design Rob Janes and Fineline Studios
Illustrator Simon Rodway

Set in Syntax and Rockwell
Colour origination by GMC Reprographics
Printed and bound in China by Hing Yip Priniting Co. Ltd

Contents

Introduction

This book is designed to guide both builders and homeowners through the potential minefield that is any building works project. What was for a long time a completely unregulated industry now has a far more defined structure, governed by a series of Acts, building regulations and local authority statutes that put the emphasis firmly on health and safety as well as quality and consistency in the building process. Nevertheless, building can still be a complex world, fraught with potential difficulties.

My earliest memories of the building industry go back to when I was a child. I remember my father getting up in the early hours of the morning and leaving for work. He was what was then known as a site agent – a person in charge of the day-to-day running of a building project. He would often take me to a site on a Saturday morning to check on progress and prepare for the early Monday start. Back then, the building industry was a very different entity from today. While construction safety booklets did exist, nobody had even thought of something like today's highly detailed Construction (Design and Management) Regulations. Nowadays, however, there is a whole raft of legislation and rules and regulations to get to grips with in the building process, which is why I have written this book.

My credentials for such a task are as follows. I started in the building industry in 1969, more than forty years ago. I worked for a family-run, medium-sized construction and civil engineering company and at the same time went to Willesden College of Technology. I have also worked as a surveyor on large construction sites as well as on small home extensions, so it is fair to say that I have a broad experience of all aspects of the building industry.

Since 1984 I have run my own surveying practice, working for the general public and local authorities as well as for private companies. I carry out surveying, estimating and project management for small to medium-sized building companies. In recent years I have prepared drawings for planning them in disputes with builders and other authorities/statutory bodies. In 2006, I was asked by a trade organization to assist them with their conciliation department and I am still doing so today.

When writing this book my aim was to provide a solid, authoritative guide to all aspects of best building practice and procedure, a guide that worked both for the jobbing builder/tradesman *and* his or her client. Hopefully I have achieved that. Simple colour coding and icons – see the key on the right – make it possible to see at a glance the information that is relevant to either party. It is full of tables, tips and illustrative examples such as sample letters. All the technical and legal information that both the builder and the client need to know is included in exhaustive yet highly accessible detail. There is advice on everything from preparing a quotation/estimate to safety on site, how to apply for planning permission and what to do in the event of a dispute between the parties involved in the project.

So, if you are looking for a straightforward, common-sense guide to how to see a building project through safely and smoothly without dispute, then look no further. Everything you need can be found in the following chapters. However, if there is one piece of advice that I would ask you to bear in mind above all others, whether you are a builder, a tradesman or a homeowner considering some building works, it would be: *remember – good communication is the key to avoid disputes and to complete the job.*

Happy reading and successful building.

Robert W. Howe

Colour coding explained:

Black text = Information for the builder/tradesman

Blue text = Information for the client

Brown text = Downloadable templates and information of particular note for the builder/tradesman or client

Icons explained:

Information and templates that Guild of Master Craftsmen members can download from www.guildmc.com

Advice aimed at the client that relates to more detailed information provided for the builder/tradesman

Advice aimed at the builder/tradesman that relates to more detailed information provided for the client

TIP CLIENT TIP Handy hints and advice for the client

BUILDER TIP Handy hints and advice for the builder/tradesman

CLIENT NOTE Information aimed primarily at the builder/tradesman, but that the client should take note of

Before You Start

Legal Requirements

In this chapter we examine the legal requirements for carrying out minor works, repairs, renovations, alterations and/or extension work to a property. This chapter briefly considers the following legislation and Acts:

- Supply of Goods and Services Act 1982
- Sale of Goods Act 1979
- Sale and Supply of Goods to Consumer Regulations 2002
- Unfair Terms in Consumer Contracts Regulation 1999
- Consumer Protection from Unfair Trading Regulations 2008.

Supply of Goods and Services Act 1982

Any person engaging a builder/tradesman to carry out work on their property will have their rights protected by the Supply of Goods and Services Act 1982. This Act makes it clear that anyone providing services must carry out the work with the following considerations:

- Reasonable care and skill
- Within a reasonable timeframe (provided a completion date has not been agreed)
- For a reasonable price (provided a price has not been agreed in advance).

Under the terms of the Supply of Goods and Services Act 1982, the builder/tradesman providing such services has a duty of care to the homeowner and their property. They should complete the work in reasonable time and at a reasonable cost, even if a price and completion date were not agreed in advance.

Sale of Goods Act 1979 (as amended)

If goods are bought from a builder/tradesman or a builder/tradesman purchases goods from his or her supplier, the Sale of Goods Act 1979 clearly states that those goods must conform to a contract and must be as follows:

- Of satisfactory quality – meaning that the product purchased should be reasonably reliable.
- Fit for purpose – meaning that the product should adequately perform the function for which it was purchased.
- As described – meaning that the product should be exactly what the builder/tradesman/supplier has described (and/or the agreed specification for the work).

Complaints can be brought to court up to six years after a sale in England and Wales and Northern Ireland and five years after discovery of the problem in Scotland. After this period, the Limitation Act 1980 generally prevents court cases being brought. This does not mean that goods have to last six years; it is not a durability requirement.

Sales receipts

It should be noted that retailers are entitled to satisfy themselves that the product was purchased at their store and on the date claimed. Producing a sales receipt as well as a credit card statement is a good way of providing such proof. Although sales receipts are not a legal requirement, purchasers are advised to request them, as they might be needed later and should, of course, be kept in a safe place.

Credit notes

Purchasers do not have to accept credit notes if goods do not confirm to the contract. However, they may be offered where the buyer has no legal right to any redress but the retailer wishes to be helpful.

The terms of the Sale of Goods Act 1979 have as much bearing on the homeowner as they do on the builder/tradesman.

If something that is purchased does not meet these standards and the goods are returned to the seller quickly, the purchaser is entitled to a refund, replacement or repair. (See overleaf for timescale of time and responsibility.)

Although the Sale and Supply of Goods to the Consumer Regulations 2002 (see page 13) pertain primarily to the builder/tradesman – as they will be supplying goods and materials to you, the homeowner – it is as well to familiarize yourself with the details of these regulations if for any reason you are not satisfied with the quality of any of the materials supplied to you.

Sale of Goods: summary of consumers' rights and remedies

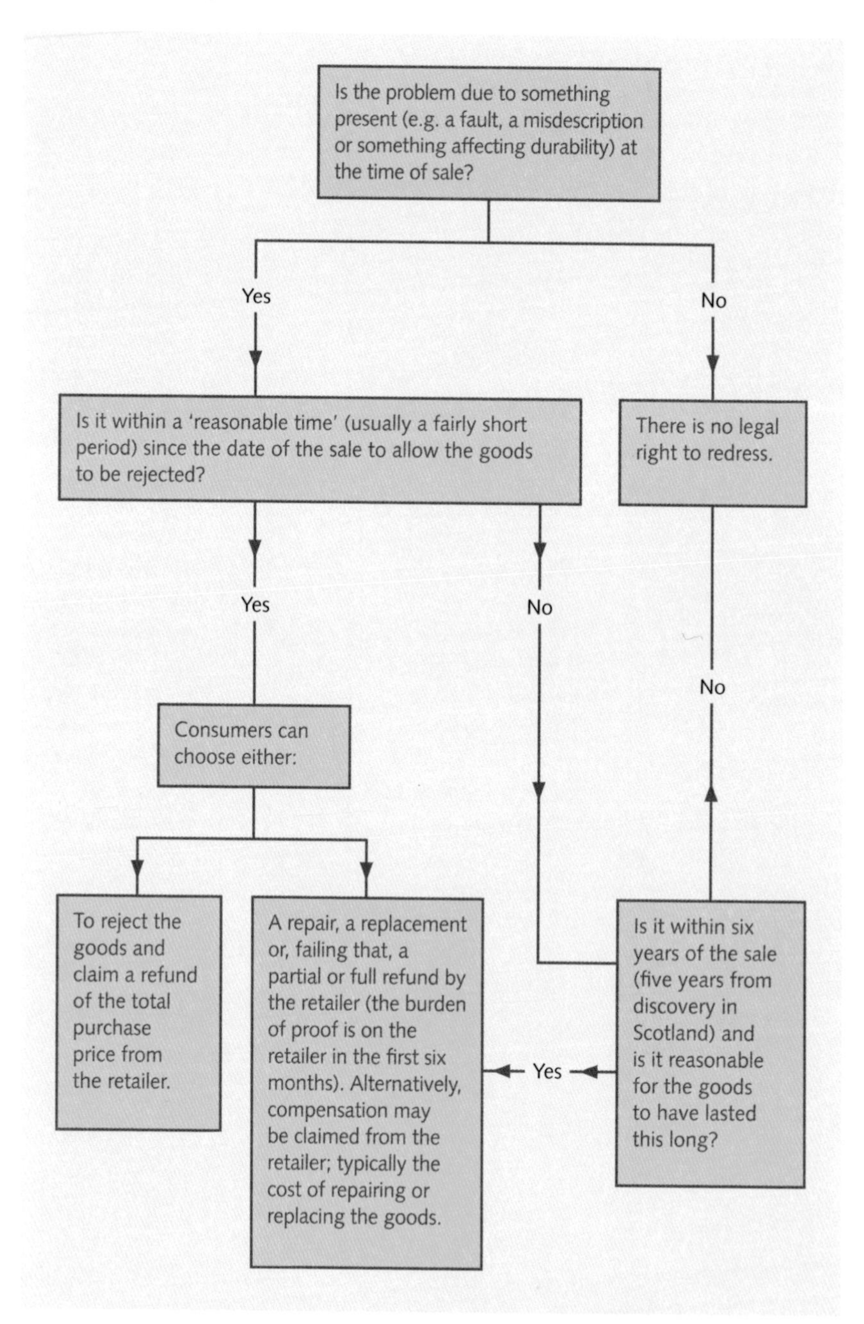

Sale and Supply of Goods to Consumer Regulations 2002

The Sale and Supply of Goods to Consumers Regulations 2002 are part of European law and are designed to create a similar standard of protection for consumers throughout the European Union.

These regulations apply to traders who sell and supply to consumers, including hire and hire purchase. The obligations to provide goods of satisfactory quality are more or less the same as those stipulated under the Sale of Goods Act 1979 (see page 11). The regulations also require the builder or tradesman to provide remedies in all cases in which goods are not of satisfactory quality.

The effect of the regulations is to create an automatic presumption – in favour of the consumer – that any defect in a product showing itself within six months is 'inherent'. This means that the defect was present at the time of sale. The burden of proof is on the trader to prove otherwise. The regulations also require that traders provide remedies whenever goods are not of satisfactory quality.

The regulations provide a time limit of six years (five years in Scotland), during which period any claims for compensation may be submitted to a court.

Unfair Terms in Consumer Contracts Regulations 1999

A consumer is not bound by a standard term in a contract with a seller or supplier if that term is unfair.

Standard terms are those devised by a business in advance and not individually negotiated with the consumer. They do not have to be provided in writing but typically are found in the small print on the back of estimates, quotations or order forms.

A standard term is unfair if it creates a significant imbalance in the parties' rights under the contract, to the detriment of the consumer, contrary to good faith.

The regulations do not cover price setting terms or terms defining the product. These are known as 'core terms'.

All terms (including core terms) must be in plain comprehensible language, as otherwise they are open to challenge as being unfair.

Consumer Protection from Unfair Trading Regulations 2008

These regulations ban traders in all sectors from using unfair commercial practices towards consumers. The regulations set out broad rules outlining when commercial practices are unfair. These fall into three main categories:

1. A general ban on conduct below a level that may be expected towards consumers.
2. Misleading practices, such as false or deceptive messages, or leaving out important information.
3. Aggressive sales techniques that use harassment, coercion or undue influence.

For a practice to be unfair under these rules, they must harm, or be likely to harm, the economic interests of the average consumer. An example is when a client makes a purchasing decision that he or she would not have made had he or she been given accurate information or not put under unfair pressure to do so.

In addition to the above, the regulations ban 31 specific practices outright.

Full details of all or any of the above Acts and regulations can be found on the following websites:

- www.dti.gov.uk
- www.oft.gov.uk

The Consumer Protection from Unfair Trading Regulations 2008 are designed primarily to protect the homeowner from unscrupulous practices by builders/tradesmen. While it is the responsibility of the builder/tradesman to operate within the terms of these regulations at all times, it is incumbent on the homeowner to call attention to any breaches as early as possible during the course of the works.

Initial Contact

For any builder, the initial contact from a potential client is very important. Making a good first impression may be the difference between landing the contract for a valuable new job or not, so it is vital to respond in a pleasant and professional manner. For the client, it is a chance to make their requirements clear from the outset.

Initial contact from a potential client could be made by any one of the following means:

- Telephone call
- Letter
- Email.

It is recommended that a ruled notebook be set aside to record all initial contacts from potential clients. Each page could be headed as follows:

Contact name	Contact date	Contact telephone number/email	Contact home/work address	Comment (e.g. work requirements, dates, etc.)

This will then give a permanent record of the initial contact, which may be subsequently referred to prior to making a first visit. It can also act as a permanent record if the initial contact puts off the visit or delays a request for an estimate or quotation.

If the initial contact is made by telephone call or email, take the opportunity to find out as much information as possible to show that you as a builder or tradesman are understanding and professional. Some initial questions to ask might be as follows:

- When would you like the work carried out?
- Is there any particular period that should be avoided, e.g. holidays etc.?
- Is there a suitable time and date that we can arrange for a visit?
- Have any local authority approvals been applied for?
- Are the neighbours aware that any work is being considered? (Getting the neighbours on board could be critical when their cooperation is necessary for access and the delivery of goods, e.g. keeping kerbs clear for delivery lorries.)
- Are any statutory authorities involved, i.e. will any moving of gas, electricity and water mains or meters be necessary? It would be wise for your prospective client to write to them as soon as possible, since statutory authorities need a long lead-in time. If the potential client is unsure, make a note to check all these services during a first visit.

Any initial contact requesting an estimate or quotation for work from a builder or tradesman should be thought through carefully; this applies whether you make the enquiry by telephone, email or letter. If you are unsure of your requirements, you cannot expect the builder or tradesman to determine them accurately.

Prior to requesting an estimate or quotation – whether it is for a cupboard or an extension – look around for as much information as possible in the following:

- Books and magazines
- Properties belonging to friends and families
- Local papers
- Websites
- DIY stores
- Architectural salvage yards (particularly if your property requires period finishes).

The more information you provide the more likely you are to receive estimates or quotations that can be compared like for like. This will then assist you in making an informed decision.

It would also be wise to state in your initial telephone call, email or letter exactly when you would like the work done, and ask how long the work is likely to take.

Do you, as the builder/tradesman, know precisely what information you require? For example, if your new enquiry is for a built-in wardrobe in a bedroom, you should ask the following:

- Is it to be full height?
- What is to be the depth and width?
- How many shelves, drawers and hangers are required?
- What materials are to be used? Medium Density Fibreboard (MDF) for a paint finish or veneered blockboard for a timber finish?
- What kind of ironmongery is required (brass, satin aluminium or chrome finish)?
- Is an internal light required?

All the above will have a cost implication and will affect the final estimate/quotation.

Having asked the relevant questions and arranged a site visit, your potential client will now consider that you are at least competent and able to carry out the work to a professional standard. This should now give you a head start on any competitors that may also be involved in this process.

In conclusion, the more information you can provide on the following:

- Style
- Materials
- Preferred time scale
- Workmanship
- Any potential problems, e.g. you know you have asbestos in your property...

... the more information the builder or tradesman has to provide you with an informed estimate or quotation. This means that you are likely to have fewer surprises when receiving your estimate or quotation and indeed when eventually the work commences.

Note: If the work you require is of such a size and value that it requires structural alteration, then it may be wise for you to consider engaging an architect or competent surveyor to prepare drawings and a detailed specification.

The Visit

A visit to the property or area of work is an opportunity for the builder to gain a better understanding of what the client requires and to discuss any potential difficulties or alternatives that may not have considered. It is also an opportunity for the client to assess whether the builder is professional in their approach.

The estimate/quotation

Visiting a client's property is an invaluable aid in the preparation of a realistic estimate/quotation. By utilizing the time spent assessing the site and the client's requirements, a realistic cost for the work can be more readily achieved. This aspect in itself will help reduce potential difficulties or disputes during the work and its acceptability on completion.

Always arrive on time and reasonably smartly dressed. If you are delayed, make sure you telephone the prospective client to check that this will be acceptable or if the appointment needs to be rescheduled.

What to look for and what to ask a prospective client

The tables on the following pages contain checklists of things that you should ask a client or look out for during a visit. By gathering all the suggested information, not only will you be able to prepare an accurate estimate/quotation, the client will see that you are organized and thorough, and therefore be more likely to give you the work.

A visit from a builder or tradesman is an opportunity for you to ask questions, to make sure they know the type and quality of work you require, and to assess the benefits of employing them.

Key questions to ask a builder or tradesman:

- Are you properly insured?
- Will your estimate include VAT?
- Will your company give a start and completion date?
- Will your company work regularly and diligently on site?

At this stage it would be wise to discuss any necessary local authority approvals, such as the following:

- Planning permission. Is this required and has the client considered or applied for this?
- Building regulation compliance. Has this been applied for? It would be wise for the client to contact their local authority for written advice.

Note: The chapter on local authority approvals (pages 56–75) looks in detail at these requirements.

It is a good idea to take digital photographs to assist in pricing before commencement of work, during the work and upon completion. This may assist in explaining to the client any variations or difficulties encountered.

An on-site discussion with a builder or tradesman will provide you with a better understanding of how costs build up within their estimate or quotation. Be aware of the difference between an estimate and a quotation, and make sure you understand what it is that you will be provided with. A quotation is a fixed price offer that cannot be changed once you have accepted it. An estimate is an educated guess at what a job may cost – but it is not binding. The price may well increase, and so you need to be sure you can cover any extra costs that may arise. (See pages 27–41 for further information.)

External work

We start with external inspection and what to look for when considering an estimate/quotation.

Parking

You may like to ask the builder or tradesman about parking arrangements for your vehicles, and if any protection will be provided, if required.

External Inspection:	Date of Visit:	
Consideration/Question	Comment/Answer	Date Actioned
1. Parking On arrival, check for ease of parking. Do you have to pay for parking? Do you need to park some distance away? *If the answer is 'yes' to either of the above a cost allowance will need to be included within the estimate/quotation or tender sum.*		

Consideration/Question	Comment/Answer	Date Actioned
2. Access Is access to the site good or poor? If you are looking at extension work, is there room to get a mechanical digger around to carry out excavations? Can a ready-mix lorry get close enough to pour the concrete into foundations and concrete ground-floor slab? Is there room to get a rubbish container on the property or will lighting and a licence be required (available from the local authority)? *There is a big difference in cost between hand dig and machine dig and in ready-mixed and hand-mix concrete. There is also a cost involved in bagging up rubbish and wheeling it some distance to a rubbish container.*		
3. Condition Check on the condition of the drive and footpath. Also check the condition of the walls and other external elements. *It is advisable to produce a schedule of condition in order that any damage that may be caused during the period of the proposed work is clearly identifiable for rectification on completion. In so doing, it will serve to counter any unwarranted claims regarding the extent of any damage arising. You may need to allow for protection or, in some instances, remove gates and railings etc. to gain access and to prevent damage. Either way, there is a cost implication to consider.*		
4. Identify drain runs This will allow you to prevent damage from heavy vehicles, rubbish containers and delivery lorries. Also, the work may include connecting to existing drains. *Protection will save extensive damage and cost involved in repairing damaged drains. Understanding existing drain runs will allow for accurate estimation of necessary work if any drainage is required.*		

Bricks

Consider how far you wish new bricks to match existing bricks. Make your requirements clear to the builder, and discuss the cost implications.

Roof tiles

For appearances, it is best to match new roof tiles to existing roof tiles. Be clear on the cost implications of this. Hand-made clay tiles, for instance, can be very expensive. Make your requirements clear to the builder.

Consideration/Question	Comment/Answer	Date Actioned
5. External materials 5.1 Bricks: If constructing an extension or just repairing external walls or porches, it is important to match the bricks as near as possible (any local builders' merchant may assist with this). There are standard bricks, such as the London Brick Company 'fletton' brick, as well as clay machine or hand-made metric or imperial-sized bricks. *Again, there is a cost implication depending on the client's choice.* 5.2 Roof tiles: If you are constructing a pitched-roof extension or porch, it's advisable to match the new roof tiles with the existing tiles (depending on the pitch of the roof). *Again, there is a cost implication between concrete tiles and clay tiles. The client should be advised and their requirements noted.*		

Window type

Think about the style of window you would like, including casement, glass and fixtures. Look at various options and consider which would work best with the style and age of your property. Discuss cost implications and your requirements with the builder.

Consideration/Question	Comment/Answer	Date Actioned
6. Window type Windows are covered under the Building Regulations (part N.) and controlled by FENSA registration. If you are building an extension or replacing a window you will need to obtain windows from a registered source and comply with the Building Regulations. Are the existing windows timber (soft wood or hard wood), uPVC or metal? Are they double-glazed? How many opening casements or fanlights are required (and consider ventilation)? Consider leaded lights (square or diamond shape), style of window, i.e. Georgian etc., type of window ironmongery, i.e. brass, gold effect or aluminium satin or silver effect. *All the above have a cost implication and should be discussed with the client and their requirements noted.*		

Consideration/Question	Comment/Answer	Date Actioned
7. Rainwater goods (gutters and downpipes) Note the style and materials of existing rainwater goods. Are they uPVC, cast iron or zinc? Are they a standard size or shape, or square or detailed? *Again, all these variations have a cost implication and should be discussed with the client and their requirements noted.* **8. Underground services** Look for service routes, particularly underground. These can be gas, electricity or water mains. It would be useful to identify gas and electricity meters as well as mains stopcocks, particularly if excavating or working on any internal pipework or electrical cabling. Try to identify telephone cabling as well. These may be overhead or underground. Do not forget to identify overhead cabling, particularly if a mobile crane is being used. *Identifying the routes of all the services and knowing where the meter boxes and stopcocks are will be invaluable if an emergency happens, such as excavating through any pipes or cabling.*		

You should never be talked into cutting corners to achieve a cheaper estimate/quote. If anything goes wrong, the fault will inevitably be blamed on you.

Internal work (condition)

One of the early aspects to consider is the overall state of the finishes internally. This will enable you to consider the extent of the making good. For example, if it is an older-style property and the ceiling is finished in lath and plaster and the requirement is to extend or form an opening for a loft hatch, it is likely that you may end up having to plasterboard and skim-coat the whole ceiling (if a dispute arises over this element of work the courts will expect you to have reasonably allowed for relevant making good after forming the opening). Again, if a new door opening is being formed in an internal wall and it is obvious on inspection that lumps of plaster are falling off, and when knocking the plaster it sounds hollow, then it is reasonable to expect that when forming the opening the majority of the plaster will fall off, so the estimate/quotation should allow for this.

It is unreasonable to price for forming the opening with minimal making good and then expect to charge extra for complete replastering of the wall afterwards.

Consideration/Question	Comment/Answer	Date Actioned
1. Internal floor and wall finishes Inspect the route through to the area in which you are going to work. What is the quality of the finishes? Are there expensive carpets, wallpaper and sanitaryware? *If so, you will need to allow for extensive protection and possibly a change of footwear, or at least cover. This will all add to the cost of the estimate/quotation.*		

Wall and floor tiling

In making good any work to wall or floor tiles, the builder is unlikely to be able to match the existing ones. You will therefore need to decide if you'd like all the tiles replaced or be prepared to find complementary tiles.

Consideration/Question	Comment/Answer	Date Actioned
2. Joinery elements What type of skirting, architraves, window sill, doors and ironmongery would the client like? Does the prospective client want you to match these in style and material? Some properties have hardwood/oak joinery, which may need to be specially made to match by a local joinery manufacturer. *If this is the case there will be a major cost implication in producing this work when compared to standard 'off-the-shelf' softwood styles. This should be pointed out to the client and, if required, two alternative costs indicated in your estimate/quotation.* If any alterations involve moving or adding items such as radiators or electrical sockets, then check pipe-work runs and cabling to ascertain the extent of the work. *If altering any gas pipework or any gas-fed boilers, fires or stoves then you will be required by law to use a Gas Safe registered engineer. If altering any electrical supplies to kitchens, bathrooms or externally, alterations must comply with part 'P' of the building regulations. It is always wise to use an NICEIC registered electrician, as, by law, all electrical work must be signed off by a certified electrician.* **3. Wall and floor tiling** If you are pricing for any making good of tiles (walls or floor) remember it is unlikely that new tiles will match the existing tiles, so give the client the option of matching as near as possible or replacing all or one section of the tiles with new (to the client's selection).		

Decorating

Decide how far you'd like the builder to go in touching up any decorations. Make sure you discuss the cost implications.

Consideration/Question	Comment/Answer	Date Actioned
4. Decorating If on completion of the work you have been asked to touch up decorations, you will need to make an assessment during the visit of just how old the existing decorations are. Any touching up with new decorations will stick out like a sore thumb on completion. Agree with the prospective client the extent of the decorations and how far to go. If you are involved in wallpapering, then the whole room will need redecorating, unless there is an adequate supply of matching wallpaper to hand. When considering decorating, assess how much preparation is required. When it comes to emulsion, does the client require matt, vinyl or silk finish? For timber does the client require varnish/stain, gloss or satinwood (a matt finish for wood) finish? If a good standard of decorating is required, it may be wise to allow for a builder's finish, i.e. prime only wood and a wash coat of emulsion (water and emulsion 1:15 mix) to newly plastered areas, making this clear in the estimate/quotation.		

The above list is not exhaustive but should give you a good idea of what to look for when making a property visit at this stage.

Conclusion

Generally, you should rely on common sense during the visit, keeping your eyes open and taking plenty of digital photographs; this will give you a better understanding of your client's requirements and assist in producing a realistically and sensibly priced estimate/quotation.

The Estimate or Quotation

In this chapter we look at the requirements for preparing an estimate or quotation. This is obviously a very important part of the process of agreeing terms for any job, which should always take place at the outset of the project.

The difference or otherwise

It is a common misconception that there is a difference between an estimate and a quotation.

An estimate is a guesstimate or an assessed figure based on minimal information and is not binding on either party. The estimate will not account for unforeseen development or work. On completion of the work you can adjust the estimate to suit your costs etc.

A quotation is a fixed sum for the work in hand. Once offered it cannot be changed. On completion, you cannot charge additional costs even if the work has changed.

The New Little Oxford dictionary defines the word 'estimate' as *'the approximate judgement of a number or value; a price quoted in advance for work'* and 'quotation' as *'quoting: passage or price quoted'*.

It is often assumed that an estimate can also be an approximate cost set against unsure criteria. However, this does not mean that you can call your price an estimate and increase the figure on completion.

It is important for you to provide as much detailed information as possible when asking for a quotation or estimate (see pages 16–17), and in either document an accurate description of the work is essential. Only genuine variations requested by you or unforeseen circumstances that could not fairly or reasonably have been foreseen can be charged as additional costs. Make sure you read the information provided here and overleaf regarding estimates and quotations so you are clear.

Let us consider the following factors:

Having been given all relevant information and made a site visit, you decide to call your proposed price an estimate with the intention of charging more on completion if you find you have underestimated the time and materials used. In the event of a dispute, the courts may determine that enough information was provided for this to be a fixed price and no additional costs can be charged.

It is only if there is a genuine variation in the work that the price may be adjusted, whether it is called an estimate or quotation.

Therefore there is very little or no difference between an estimate or a quotation. It is the information provided and the description of work to be carried out that will determine the contract between both parties and not the words 'estimate' or 'quotation'.

Preparing the estimate or quotation

Having received your initial enquiry and made a visit to the premises where the work is to be carried out, you are now ready to sit down and prepare your estimate or quotation.

For the purpose of this exercise we will prepare an estimate for a small, rear single-storey kitchen extension, size 4m wide x 3m deep in which the rear wall of the existing kitchen will need to be propped and removed and a universal beam installed to carry the rear first-floor external cavity wall. The client also requires new kitchen units to be supplied and installed with new worktops, floor and wall tiles and for the extension to be painted inside and out.

The client has made available detailed drawings prepared for planning permission and building regulation approvals.

We have also assumed that Construction (Design and Management) Regulations 2007 and The Party Wall etc. Act 1996 do not apply. These regulations and Act will be covered on pages 76–91.

We are assuming a contract period of eight weeks. The following items could be included in the estimate or quotation:

- Preliminaries
- Provisional sums
- Prime cost (P.C.) sums
- Elemental breakdown of the work, to include costing of labour and materials
- Overheads and profit margins.

The above items are explained as follows:

1. Preliminaries

When preparing your estimate or quotation there are certain items of expenditure that are covered under the heading of 'Preliminaries'. These are items that generally cannot be satisfactorily split between the elemental costs. Examples of these are as follows:

Site supervision: Non-productive time spent supervising the work, speaking to the householder, etc. This could be listed as so many hours per week over the period of the contract.

Example: Say three hours per week over an eight-week contract period giving 24 hours of non-productive time, assuming an hourly rate of £18.50, this would give a **cost of £444.00**.

Waste containers for rubbish: It is important to keep a clean site and it is up to the individual builder to estimate the number of skips to be used.

Example: Assume that seven skips are used and that their cost is £165.00 per skip – this would give a **cost of £1,155.00**.

Access scaffold: A scaffold company will charge for erection and dismantling of scaffolding.

Example: This may be in the region of say £800.00 plus a monthly hire charge of, say, £135.00 per month – therefore a figure of £800.00 + (2 x £135.00) £270.00 would give a total **cost of £1,070.00**.

Small tools: This item is for the wear and tear and eventual replacement of small tools such as drills, grinders, saws, chisels and levels, etc. It is advisable to cost out such tools by assessing the cost of buying these tools; in this instance, we are basing our sums on £1,000 and we assume that the tools will last two years before they need replacement.

Example: If we divide £1,000 by 104 weeks, this gives a weekly figure of, say, £9.60 per week. Therefore, for a contract period of eight weeks, we arrive at **a cost of £76.92**.

Insurance for the work: If your public indemnity insurance and employers' liability insurance and any other insurance you may have, such as contractors all risk insurance, totals, say, £952.50 per annum, you need to divide the total cost by 52 weeks (if there is more than one contract, the weekly figure should be broken down pro rata).

Example: An annual cost of £952.50 should be divided by 52 and multiplied by 8 to give **a cost of £146.53**.

Toilet facilities on site: Very few clients encourage contractors to use their bathrooms; therefore separate facilities are almost always required. Portable toilets are usually available for a set weekly sum that includes emptying and replenishing of necessary cleaning fluids.

Example: A weekly hire charge of £30.00 for eight weeks would give **a cost of £240.00**.

Telephone for the work: Mobile phones are now almost always used exclusively on small to medium contracts. Therefore a pro rata figure should be calculated for the monthly connection charge and a proportion of the calls.

Example: If your monthly connection charge is £27.00 plus call costs of say £45.00 per month then the cost per month = £72.00. Divide this by 4.34 = £16.59 per week x an 8-week contract period to give **a cost of £132.72**

On occasion, a builder or tradesman might ask if they can use one of your personal small tools, if they are not in possession of exactly what they need at a particular time. This is generally not advisable and should be avoided with a polite refusal in most instances. This is because your tools could be spoilt during the work, or mistakenly included in the builder's tool box and removed from the premises.

It is entirely your decision as to whether you choose to let a builder or tradesmen use your toilet facilities or not. Most clients choose not to, but sometimes there are issues of space on site that preclude the builder/tradesman from installing their own temporary facilities. As always, clear communication about this matter at the outset of the job will reduce the likelihood of problems later on.

Although it should not generally be necessary, it might be a good idea to explain tactfully at the outset that it is not acceptable for the builder/tradesman to use your home telephone landline at any time. All builders/tradesmen carry mobile phones, so there should be no excuse when it comes to ordering extra materials etc. on site.

Protection: Allow for a cost for temporary protection when the external wall is removed. Allow to tape up doors in an attempt to stop dust from spreading. Temporary protection to carpets and other finishes as required. These will need to include the cost of ply and 75 x 50mm softwood studs to opening.

Example: To provide a temporary partition would require three sheets of 18mm ply at £33.00 per sheet = £99.00 plus 75 x 50mm softwood studs, say, 30.6m at £2.04 per m = say, £62.40 plus a carpenter for, say, five hours at £18.75 per hour = £93.75. Therefore, the total cost of the temporary partition would come to £255.15. To provide temporary dust sheets, protection and tape for sealing doors = 30 minutes per day x five days per week = 2.5 hours per week x £15.00 for general labour to protect at start of day and remove at end of each day = £37.50 x eight weeks = £300.00 plus materials, say, three dust sheets and roll of tape, say = £68.50 = £368.50. Therefore protection for a contract of eight weeks would **cost £623.65**.

Whilst this list is not exhaustive, these are a few of the important items that should be included in your preliminaries section for this type of work.

In this example, the total cost of preliminaries carried forward to summary = £3,888.82.

This figure can be used as a guide for your weekly preliminary costs, i.e. £3,888.82 divided by eight weeks = £486.10 per week. Should your client ask you to carry out additional work or varies the work in some way leading to additional time, there is now a method of claiming for additional cost of preliminaries.

For example, if the additional and varied work has led to the contract period being extended from eight weeks to ten weeks, there is a two-week extended contract period. If this extended period can be clearly identified, then the weekly cost of the preliminary items can be worked out and multiplied by two to give an additional cost to be added onto the final account. Or, as often happens, the additional time is two days here and two days there; by adding up the days you will be able to arrive at an assessment of the additional period with each week (in this particular example) being chargeable at £486.10 or pro rata in your final account. It must be pointed out that this is an assessment and not an actual cost, so this method should only be used to negotiate a figure for inclusion in the final account. Record keeping is essential, as it will assist in identifying the delays and additional preliminary costs.

2. Provisional sums

A provisional sum is a sum allowed for an element of work that cannot be entirely foreseen, or cannot be defined or detailed at the time of preparing the estimate or quotation.

For this example, we may use the new kitchen units and worktops. We understand that the client has seen some kitchen units in a local joinery supplier. In this case, we would allow a provisional sum of £2,500.00. When it comes to buying the kitchen units and worktops, an accurate record of their cost along with a copy of the invoice should be attached to the final account. Time sheets for fixing them should be kept and copied into the final account.

Let us assume the kitchen units and worktops cost £1,350.00. You should add a percentage for profit of, say, 20 per cent* = £270.00 giving a total of £1,620.00 plus, say, three days fixing for tradesman and labourer assisting. Therefore three days x eight hours = 24 hours x (£18.75 + £15.00) £33.75 = £810.00. Therefore, the cost of kitchen units in this instance is **£2,430.00**.

If your estimate or quotation included the provisional sum of £2,500.00 then this sum should be omitted and the actual cost, including profit, £2,430.00 should be added back. In this example the property owner would have saved a sum of £70.00.

3. Prime cost (P.C.) sums

This is a sum provided for work or services to be executed by a sub-contractor nominated by the client or a statutory supplier. Such sums shall be exclusive of any profit, and percentage profit shall be included and identified in the final account.

An example may be that the gas supplier needs to give an estimate to move the gas meter. Let us assume they quote £300.00 for moving this meter: then this sum should be included in the estimate or quotation with an identifiable percentage for overheads (say, 15 per cent and, say, percentage profit of 20 per cent*). The sum to be included in the estimate or quotation should be £300.00 + £45.00 + £60.00.

* These percentages are only an example for the purpose of this exercise.

4. Elemental estimate or quotation

This section is where the estimate/quotation is broken down into the following elements:

I. Site set-up		£ 645.00
II. Demolition and removals		£ 566.00
III. Structural alterations		£ 1,315.00
IV. Foundation		£ 695.00
V. Concrete work		£ 985.75
VI. Brickwork and blockwork		
(a) Up to damp-proof course:	£ 855.35	
(b) Above damp-proof course:	£ 2,993.35	
		£ 3,848.70
VII. Roofing work		£ 1,340.50
VIII. Carpentry and joinery		
(a) Roof	£ 645.50	
(b) Soffit and fascia boards	£ 327.80	
(c) Boxing in	£ 290.50	
(d) Windows and doors	£ 1599.20	
(e) Skirting and architraves	£ 637.00	
(f) Door sets	£ 495.00	
		£ 3,995.00
IX. Plasterwork and other finishes		
(a) Plasterwork	£ 604.65	
(b) Screed	£ 596.75	
(c) Wall and floor tiling	£ 665.50	
		£ 1,866.90
X. Services		
(a) Plumbing work	£ 1270.00	
(b) Electrical work	£ 710.00	
		£ 1,980.00
XI. Painting and decorating		£ 1,375.00
XII. Drainage		£ 669.25
XIII. External work		£ 890.00
Total to summary page		**£20,172.10**

The aforementioned sums are assessed for the purpose of this exercise (with the exception of parts of the brickwork above DPC) to show how your estimate can be built up.

These headings can also be used to formulate a programme of works as well as a schedule of payments. The chapter on Programme and Progress of the Work (see pages 110–121) will explain how this can work.

The headings indicate the element of work. On the facing page is an indication of how to build up a rate for one element of work, in this instance section VI Brickwork and blockwork, which can be used as a guide for most other elements of work.

Here we look at the cost of the external cavity walling to the extension.

We start with calculating the area of brickwork required, as follows: The length of the back wall is 4m x its height of, say, 2.25m from DPC = $9m^2$. The depth of the two side walls is 3m x its height of 2.25 from DPC = 6.75 x 2 = $13.5m^2$. Therefore 9 + 13.5 = $22.5m^2$. Deduct door and window openings say = $4.99m^2$ to give a total area of cavity brick and blockwork above damp-proof course of $17.51m^2$.

In the event of a dispute, the courts will require the actual cost of any disputed work or goods to be included and not general assessments, unless you can prove that this is a reasonable way of identifying cost.

We continue with calculating the cost of laying a metre square of brickwork, as follows:

Description	Area of brickwork	Cost of bricks	Labour	Plant	Total rate
Half brick wall in skin of cavity wall 300mm wide. In facing bricks at £450.00 per 1000.	As set out above.	No. of bricks = 17.51 x 60 bricks in a square metre = 1051 plus 15 per cent wastage of 158 giving a total of 1209 bricks x £450 per 1000 = £544.05 plus, say, £135.00 for sand and cement and any required additive = £679.05 divided by 17.51 = rate for cost of bricks is:	It should take a 2+1 gang (i.e. two bricklayers and a labourer) to lay 1209 bricks (at 600 per day) = 2.02 days. Bricklayer at £150 per day. Labourer at £120 day. This gives a daily rate of £420 a day x 2.02 days = £848.40 divided by 17.51 is:	Three days hire for mixer could be, say, £22.50 per day = £67.50. Grinder for three days at £11.25 per day = £33.75 Total cost of plant hire = £101.25 divided by 17.51 is:	Therefore the total rate for a half brick wall in cavity construction built in facing bricks costing £450 per 1000 is:
	$17.51m^2$	$£38.78m^2$	$£48.45m^2$	$£5.78m^2$	$£93.01m^2$

We continue calculating as follows:

Half brick wall in skin of cavity wall, 300mm wide. Built-in facing bricks at £450.00 per 1000

17.51m^2 @ £93.01 = £ 1,628.61

100mm insulation block in skin of hollow wall

17.51m^2 @ say £44.98 = £ 787.60

Cavity construction including wall ties and 100mm insulation

17.51m^2 @ say £11.67 = £ 204.34

Stainless steel vertical connection of new brick to existing

8.8m @ say £17.68 = £ 155.58

UPVC insulated cavity closer

6.6m @ say £18.98 = £ 125.27

Steel insulated lintel for 300mm cavity wall length 2.100mm.

1 no. @ say £ 54.80 = £ 54.80

Ditto to 1500mm

1 no. @ say £ 37.15 = £ 37.15

The total cost of brickwork above damp-proof course (DPC) to be carried forward to summary page = £2,993.35

In addition to the above

Total cost of brickwork bel DPC c/f to summary page = £855.35

The rates used in 'Half brick wall' in cavity construction can change depending on the laying rate of individual gangs and the complexity of the work at the time. For example, if the work is two or three storeys up, then you need to make allowance for time to offload bricks and hoist or cart them up to the second or third floor.

This is only one element of the brickwork section that will eventually build up to your cost for insertion to the brickwork section of the summary sheet. For demonstration purposes, sample items and costs have been inserted to allow sums to be forwarded to the example elemental cost table.

5. Overheads and profit margins

If estimates or quotations were produced based purely on site costs, financially they could lead to a loss situation. The hidden costs of running a business are known as the 'overheads'.

Overheads

To arrive at an accurate estimate or quotation you need to include a sum for overheads. Items included under the heading of overheads are as follows:

Salaries for office working
10 hours per week x 48 = 480 x £18.75 = £ 9,000.00

Office costs working from room at home, say, yearly rates £1,700.00, say, one room = seventh of rates = £ 242.86

Motor vehicle expenses for van or truck, say, monthly purchase cost = £199.00 x 12 = £2,388.00 plus diesel at, say, £120.00/month x 12 = £1,440.00
Servicing cost, say, £300.00.
Total cost = £ 4,128.00

Printing and stationery allow a sum of £ 300.00

Advertising allow a sum of £ 875.00

Postage and telephone allow a sum of £ 750.00

Professional fees, accountants, etc.
allow a sum of £ 600.00

Bank charges overdraft set-up cost on borrowing of £5,000.00 = £450.00 plus monthly charges at £70.00 x 12 = £840.00 = £ 1,290.00

Yearly overhead costs = £17,185.86

In our example we have assumed a company turning over £175,000.00. Again, assessed figures above have been included to indicate how a reasonable percentage can be added on a job-by-job basis. Therefore, looking at the above example, it can be calculated that the yearly overhead cost of £17,185.86 as a percentage of the annual turnover of £175,000.00 = 9.82 per cent.

Profit

It should be understood that a business is a commercial enterprise and, as such, should make a reasonable profit over and above the directors' or sole proprietor's fees paid for their services. The profit can be used to pay the directors a dividend or the sole proprietor a bonus over and above their wages.

There is no right percentage for a profit margin, as a large company with a multi-million pound turnover has profit margins as low as one or two per cent. However, many small businesses can work with a profit margin of between 20 to 50 per cent, depending on the degree of difficulty and risk. This has to be assessed by the directors or sole proprietor on commercial grounds.

The profit margin to be included will be influenced by several factors as follows:

- Work in hand
- Future commitments
- How competitive the recent estimates or quotations are
- How much future work may be received from this source
- How difficult this work is and what degree of risk is involved.

Assessment of profit percentage

Using the current example:

- The contract is a reasonably simple job with only a small amount of risk (insertion of two universal beams carrying external cavity first floor walls)
- The contract period is fairly short
- Turnover is difficult at the moment
- The property owner has discussed the possibility of further internal alteration work (useful during the winter period).

In this fictitious situation it is assessed that the profit margin should be 20 per cent.

Below is an example of a summary page for this fictitious estimate or quotation for a kitchen extension and internal alterations.

Summary page

1. Preliminaries =	£ 3,888.82
2. Provisional sum for the supply and fit of kitchen units and worktops =	£ 2,500.00
3. Prime cost sum for work by gas supplier to move meter to new position =	£ 300.00
4. Elemental work section =	£ 20,172.10
Sub-total =	£ 26,860.92
5i. Overheads @ 9.82 per cent =	£ 2,637.74
5ii. Profit margin @ 20 per cent =	£ 5,372.18
Total carried forward to estimate/quotation =	**£ 34,870.84**

Plus necessary Value Added Tax (VAT) at the present rate (currently 17.5 per cent; as of January 2011 VAT increases to 20 per cent)

All of the above information indicates how to provide an accurate estimate or quotation for the works described. The principles remain the same when providing an estimate or quotation for any work around the property. Only the percentages will alter, depending on the size of the required work.

The following letter is an example of how to present an estimate or quotation in letter form to the prospective client.

Our ref: Est/quot. 435 dd/mm/yyyy

Mr Client
Address
County
Postcode

Dear Mr Client,

Re: Estimate for proposed kitchen extension and internal alteration at address.

We would like to thank you both for the taking the time to explain your requirements and show the undersigned around your property during his/her recent visit, further to which we are pleased to provide our estimate (or quotation) as follows:

To construct single-storey rear extension and internal alteration to form an extended kitchen and breakfast area, all as drawings and documents provided as follows:

Drawing number PC/001/rev B
Drawing number PC/002/rev A
Drawing number PC/003/rev C
Drawing number PC/004/rev B

Brief specification of work provided by A.N. Architect and dated 22nd March 2010.

Based on our site visit dated 7th April 2010.

All for the sum of **£34,870.84**
Plus VAT at the present rate of 17.5%
(VAT due to increase to 20% in January 2011)

Based on continuous working, the length of programme allowed is eight weeks, subject to the kitchen supplier and the company nominated by the gas supplier being able to work to the time period set out in the programme.

In order for the above sum to be realistic, we have included the following criteria:

1. Foundations to 1000mm depth filled with lean mix concrete.
2. Cavity external walls in half brick wall external skin and 100mm insulating block inner skin. Cavity filled with 100mm insulation.
3. 100mm concrete slab laid on 150mm well-compacted hard core with 50mm sand blinding with 1200g D.P.M.
4. 75mm sand and cement screed laid on 100mm flooring grade insulation.
5. Pitched roof construction to match pitch of existing roof and including matching roof tiles.
6. Windows and doors to be white uPVC double glazed.
7. Remove existing cavity wall and install universal beams to form extended kitchen.
8. Plaster all internal walls and ceilings.
9. Allow the provisional sum of £2,500.00 for supply and install kitchen units and worktops.
10. Allow the P.C. sum of £300.00 for the moving of the gas meter.
11. Allow for floor and wall tiles as indicated.
12. Allow for internal decorations to new plastered areas.
13. External work allow for side paved path to give access from front to back.
14. On completion leave the property clean and tidy internally and externally.

This estimate or quotation excludes the following:

1. Any local authority and professional fees.
2. Any work outside the kitchen area.
3. Any external work unless described above.
4. No allowance has been made for Construction (Design and Management) Regulations 2007.
5. No allowance has been made for complying with the Party wall etc. Act 1996.

We hope that this estimate (or quotation) meets with your approval. If we can be of any further assistance please do not hesitate to contact us and we will be happy to answer any further questions or discuss any other requirements.

Yours sincerely,

Name
For company

Negotiating the Contract

Once the works have been awarded it would be wise to negotiate and agree a formal contract. These are usually available from trades organizations or can be purchased from the Royal Institution of Chartered Surveyors or from the Royal Institute of British Architects.

For work around the home and for small extensions, a small works contract would be suitable. The contract needs to be clear and understandable and fair to both parties.

If a problem arises then a look in the contract at the relevant clause should give clear advice as to who is responsible and how to overcome the problem.

It is important to include fair and reasonable information in a contract. This means, for example, that such items as start and completion dates must be realistic and not just randomly inserted to please one party or the other.

- Insert a clear contract sum (excluding or including VAT)
- Set out what the contract sum was based on, i.e. list the contract documents
- Set out how to deal with variations and delays

An example of a standard small works contract is shown on pages 43–48.

The contract should be clear and understandable. If there are any elements within the contract that you don't fully understand, ask your builder or tradesman to clarify.

Building Agreement for Minor Building Work

THE PARTIES

Builder/tradesman:

Name of company or sole trader: ______________________________

Address: ______________________________

____

____

Phone number: ______________________________

Email address: ______________________________

VAT registration number: ______________________________

If not registered for VAT confirm here: ______________________________

Customer:

Name: ______________________________

Address: ______________________________

____

____

Phone number: ______________________________

Email address: ______________________________

Address where work is to be carried out ('premises'), if different to that above:

____

____

____

Agreement on the work to be carried out:

1. This agreement shall incorporate those conditions set out in Schedule 1.

2. Agreed work (A short description):

____

____

____

____

____

____

3. List documents used to achieve the quotation:

3.1 Builder's/tradesman's quotation
Date of quotation: ______________________________

3.2 Drawings used: ______________________________

3.3 Drawing numbers: ______________________________
____
____
____

3.4 Specification used: ______________________________

3.5 Name and date of specification: ______________________________
____
____

3.6 Other documents used, e.g. material or product supplier's specification:
____
____
____
____

4. Agreed price for the work:

The agreed price for the work described above all for the sum of:

£ ______________________________

Inclusive of VAT, and anything that can reasonably be expected from carrying out a careful inspection on the pre-estimate/quotation visit.

Any agreed exclusions: ______________________________
____
____
____

5. Local authority approvals and other approvals required:

5.1 Planning permission, building regulations, Party Wall etc. Act 1996 and Construction (Design and Management) Regulations 2007.

5.2 Confirm here that planning permission, building regulation and Party Wall etc. Act 1996, if required, has been applied for and the relevant permissions received:

(a) Planning permission application no.: ______________________
Planning permission approved date: ______________________

(b) Building regulation application no.: ______________________
Building regulation notice of approval date: ______________________

(c) The Party Wall etc. Act 1996 Date neighbour notified: ______________
Date inspection carried out: ______________

(d) Construction (Design Management) Regulations 2007.

6. No work is to start until all of the above has been considered and relevant permissions received. The builder/tradesman can start work before building regulation approvals have been received provided an application number has been obtained and the local authority has been given 48 hours' notice.

7. The customer allows the use of the following facilities:

Electricity: ______________________
Water: ______________________
Telephone: ______________________
Toilet: ______________________

The use by the builder/tradesman of any or all of the facilities listed in clause 7 of the sample contract (see above) is entirely at your discretion as the homeowner. If you don't want your facilities used, then say so and insist that this clause is omitted from the contract.

With regard to clause 8 of the sample contract (see below), bear in mind that it is your responsibility as the customer to ensure that any variation in the work is properly notified in good time, in order to achieve an appropriate increase or reduction in the agreed price.

8. Variation in the work: If there is any variation in the work requested by the customer or the local authority that could not reasonably have been foreseen, then the agreed price will be increased or reduced depending on the variation by an amount or time period (as the case may be), to be agreed between the parties.

9. Payment (by one of three options):

Option A [used/not used]

9.1 The agreed price should be paid on completion of the agreed work.
£ ______________________

Option B [used/not used]

9.2 The agreed price shall be paid by stage payments agreed at the outset as follows:

Stage:	Payment:
__________	£ __________
__________	£ __________
__________	£ __________
__________	£ __________

Option C [used/not used]

9.3 The agreed price shall be paid, based on work carried out to date at [ ] intervals.

9.4 Irrespective of which pricing option is chosen, the customer shall be entitled to deduct from each payment to allow a 95 per cent payment on completion of the agreed works, which should be paid within seven days of the above agreed dates or stages.

9.5 The remaining five per cent of the agreed price should be paid within one month of completion of the agreed works, provided that all matters at issue have been dealt with satisfactorily.

10. Programme of work:
(If work is to last more than four weeks, a programme of work will be issued)

10.1 The work will be completed within ________weeks of the agreed start date.

10.2 The work will commence on: ______________________________

10.3 The work will be completed by: ______________________________

10.4 If the customer issues any late instructions or there is extremely inclement weather, the builder/tradesman shall be entitled to an extension of time to the programme to be agreed with the customer and which may have an effect on the contract period. For the avoidance of doubt, the occurrence of extremely inclement weather shall entitle the builder/tradesman to an addition to the programme set out in this clause.

11. Insurance: Before commencement of the agreed work the customer will inform his/her insurers of the work to be carried out.

11.1 The builder/tradesman will have public liability insurance.

11.2 The builder/tradesman will have employer's liability insurance if relevant.

11.3 The builder/tradesman will also insure the customer against theft of any materials or products that are for the agreed work.

12. Guarantees/warranties: All product or material guarantees issued by the manufacturer must be handed over to the customer before the final five per cent retention sum is handed over.

13. Security issues: The customer will inform the builder/tradesman in writing prior to commencement if the premises will be occupied throughout the work. The builder/tradesman will at all times make reasonable efforts to keep the premises secure.

You should take careful note of the following clauses in the sample contract, as these have a significant bearing on your rights as a customer: 10.4; 11; 11.3; 12; 13 (see page 47). If you are unsure of the meaning of any of these terms, you are advised to consult your solicitor.

14. Disputes:

14.1 Both parties should endeavour to use a mediation or concilliation service prior to starting any court proceedings.

14.2 The customer or the builder/tradesman can commence court proceedings to settle any disputes.

THIS AGREEMENT IS MADE BETWEEN:

Builder's/tradesman's signature: ______________________________

Date: ______________________________

Customer's signature: ______________________________

Date: ______________________________

Following on from the contract, it is advisable to include a Schedule of Conditions, which indicates responsibilities and provides further explanation of the clauses included within the contract.
A typical example is as follows:

All the information given in Schedule 1 below is relevant to the homeowner as well as the builder/tradesman and should be read thoroughly by both parties.

SCHEDULE 1
THE CONDITIONS

1. Builder's/tradesman's responsibilities:

1.1 The builder/tradesman shall carry out the agreed work in a proper and workmanlike manner.

1.2 Use materials that are of a satisfactory quality and suitable for their intended purpose. The materials will be new unless otherwise agreed with the customer. The builder/tradesman shall also not include in the agreed works any materials that are generally known in the industry to be harmful.

1.3 Work regularly and diligently. Complete the work as per the agreed programme or according to any extension of time agreed between the parties.

1.4 Agree any requested variation of cost with the customer in writing before conducting any such works.

1.5 Keep reasonable written records of variation work and their effect.

1.6 Agree at the outset any work to be sub-contracted.

1.7 Leave the property clean and tidy at the end of each and every day's work.

1.8 Dispose of any rubbish or unwanted materials from the premises.

1.9 Be responsible for any damage caused to the premises and its contents and the same to neighbouring properties.

1.10 Keep the premises secure at all times.

1.11 Keep noise to a reasonable level.

1.12 Treat the client and their premises with respect.

2. Customer's responsibilities:

2.1 Give the builder/tradesman access to the premises during the agreed programme period.

2.2 Remove any furniture and objects within the working area to allow the builder/tradesman reasonable access.

2.3 Allow the builder/tradesman to carry out the work in any sequence required to complete the work.

2.4 Treat the builder/tradesman with respect.

2.5 Pay the builder/tradesman as stated in the agreement.

2.6 Confirm in writing any variation to the work.

3. Health and safety:

The builder/tradesman must:

3.1 Consider whether Construction (Design and Management) Regulations 2007 apply and, if so, comply with any requirement of those regulations.

3.2 Prevent or minimize health and safety risk to the customer, any workforce and others that might be entering and leaving the premises.

3.3 Comply with environmental and pollution regulations and requirements.

3.4 Make sure any temporary protection for the agreed work is safe and weatherproof.

3.5 Protect the premises at all times from access by visitors or children.

The customer must:

3.6 Take notice of instructions or information put up by the builder/tradesman.

4. Extending the agreed programme:

An extension to the agreed period will be agreed by a fair and reasonable time if the builder/tradesman:

4.1 Has to spend additional time on the work due to changes made by the customer or local authority.

4.2 Cannot complete the work on time because of reasons beyond the builder's/tradesman's control, such as late or incomplete decisions made by the customer or extremely inclement weather.

5. Completion:

5.1 On completion of the agreed work and following a joint inspection between the customer and the builder/tradesman, a reasonable snagging list will be issued. The builder/tradesman then has one month to complete all snagging issues and hand over any guarantees and/or warranties before final payment is made.

5.2 Apart from guarantees and/or warranties, the contractor remains responsible for faulty workmanship and or materials used for up to six years.

6. Termination of this agreement:

6.1 The customer may terminate the agreement if the builder/tradesman:

(a) Does not work regularly and diligently.

(b) Is not meeting health and safety, environmental and/or pollution responsibilities.

(c) Does not work competently or is careless leading to an unacceptable standard.

(d) Does not correct the work or situation within seven days of receiving a written warning by the customer.

(e) Is the subject of insolvency or bankruptcy proceedings.

The customer will need only to pay a reasonable sum for work carried out and materials on site that are to a reasonable standard.

6.2 The builder/tradesman may terminate the agreement if the customer:

(a) Fails to make a payment without fair and reasonable justification.

(b) Stops or obstructs the work from being carried out and does not correct matters within seven days of receiving a written notice from the builder/tradesman, then the builder/tradesman may end this agreement by giving a final written notice.

(c) Is the subject of insolvency or bankruptcy proceedings.

(d) If the builder/tradesman ends this agreement, then the customer will within ten working days make a fair and reasonable payment for work and materials fairly and reasonably carried out.

6.3 The customer and the builder/tradesman can claim from each other for fair costs and expenses that resulted from either side failing to keep to the terms of this agreement.

Also included in most contracts are guidance notes.
A typical example is as follows:

BUILDING AGREEMENT FOR MINOR BUILDING WORK (the 'Agreement')

Guidance notes

1. The Agreement

1.1 Parties
The builder/tradesman should ensure that the relevant names, addresses and contact details are stipulated in the relevant positions on the front page. If the builder/tradesman, or indeed customer, is not an individual but rather a limited or otherwise registered company, the name of the company should

be inserted, as should its registered office. The builder/tradesman should also insert their VAT registration number or confirm if not registered for VAT purposes. The builder/tradesman should also stipulate the premises or site at the place indicated at the bottom of page 1 where such property is different to the address used in the customer's details.

1.2 Clause 2: Agreed works
The builder/tradesman should include a short description of the work that has been agreed and the subject of the quotation or agreed price for the work. The description need not be excessive, but should adequately describe the work to be done and highlight any specific issues such as design that the builder/tradesman may need to undertake.

1.3 Clause 3: List documents used to achieve the quotation
The builder/tradesman should include, to the extent the same exist, the details of any drawings or specifications or other documentation used in relation to the works and as specified in their original quotation.

1.4 Clause 4: Agreed price for the work
The builder/tradesman should insert the agreed lump sum price in respect of the agreed works above. The builder/tradesman should note that this figure is quoted inclusive of VAT and so the builder/tradesman will need to account for their VAT liability to Her Majesty's Customs & Revenue. The builder/tradesman should also note that the agreed lump sum will also include any costs that could reasonably be expected from carrying out an inspection on the pre-estimate or quotation visit. The builder/tradesman should also list any agreed exclusions from the agreed price for work for which they may receive additional monies.

1.5 Clause 5: Local authority approvals and other approvals required
Where the builder/tradesman is aware of any building permissions or applications for building regulation approval, they should insert the details at the relevant positions in Clause 5.2. The builder/tradesman should also insert in relation to any Party Wall or Construction (Design and Management) Regulations 2007 (CDM) notifications or issues of which they are aware at 5.2(c). The builder/tradesman should also note additional CDM obligations if a project is notifiable, i.e. longer than 30 days or more than 500 person days.

Pay close attention to 1.6 Clause 7: Utilities in the sample contract guidance notes (see below).
A builder/tradesman only has a right of access to your facilities with your express permission.

1.6 Clause 7: Utilities

The builder/tradesman should confirm with the customer which utilities or facilities as specified in clause 7 the builder/tradesman will be able to use. The builder/tradesman should note that if the customer does not allow the use of the facilities indicated the builder/tradesman should incorporate into the agreed price for the works an element of cost to cover any rental or utilities bills that they may incur as a result.

1.7 Clause 9: Payment

The payment provisions at Clause 9 allow for three options to be used.
Option A: For payment of a lump sum on completion.
Option B: For payment by stage payments.
Option C: Payment at periodic intervals based on work carried out to date.
The builder/tradesman should delete the reference to 'not used' for the option they have chosen for the pricing and in relation to the other two options under Clause 9 they should delete 'used'.

The builder/tradesman should note that for commercial projects with an expected bill time of over 45 days, Option B should be used as this would meet the requirements of the Housing Grants Construction and Regeneration Act 1996.

1.8 Clause 10: Programme

The builder/tradesman should take care to complete section 10 appropriately if the work is to last more than four weeks. An appropriate programme should be inserted and the builder/tradesman should be mindful that the programme may be subject to extensions of time in certain instances, as specified both here and in the conditions.

1.9 Clause 11: Insurance
The builder/tradesman should agree in advance with the customer what insurance they will take out and maintain and also the level of such insurance. These relevant figures and the basis for insurance should be inserted behind the colon in Clause 11.1 (for public liability insurance) and Clause 11.2 for (employer's liability insurance).

1.10 Execution/signature
As currently drafted, this minor form of contract allows for execution underhand, that is execution and dating by the builder/tradesman with their signature. Where the works may be more substantial, the builder/tradesman and customer should consider whether execution as a deed is preferable. If it was decided, or indeed requested by the customer, that the contract is executed by a deed, then the builder/tradesman could do so by signing the document in front of a witness who would also have to sign and insert their name, address and occupation details. The customer could similarly execute. An alternate form of execution in which either the builder/tradesman or the customer may be a limited liability company is for a director to execute in the same way, or for two directors or a director and company secretary to execute.

While the above contract, conditions and general guidance notes may appear to be a lot to take in, they do give all parties the opportunity of asking relevant questions and clearing up any discrepancies prior to work starting, as well as showing an orderly procedure in which to resolve any possible disputes.

If a client wants to include a damages sum for late completion of the agreed works, you would be within your rights to claim for an additional amount equal to the sum mentioned if you were to complete early. This of course must be agreed and included within the contract.

Penalty clauses are unenforceable in law. Therefore, any costs you may establish for delay have to be easily identifiable as genuine damages and, as such, be a pre-estimate of costs for delay.

Local Authority Approvals

The planning system plays an important role in helping to protect the environment in our towns and cities as well as in the countryside. Each local planning authority must produce a local development framework that outlines how planning will be managed in their area. The overall objective of this chapter is to improve an understanding of local authority approvals, while reinforcing information that builders and tradesmen use every day in their working life.

Please note that the information provided in this chapter is a general guide for possible work being undertaken and is not a definitive source of legal information. It is recommended that the local authority is contacted prior to any work being carried out to any property.

This chapter is primarily aimed at your client, as in the majority of cases the property owner will apply for planning permission and other local authority approvals. However, much of the information herein is key to the entire building process – especially building regulations – and should be duly noted and acted upon by you.

The local planning authority is responsible for deciding if a development, whether it be an an extension to a residence or a new shopping centre, will receive the necessary permission.

There is an appeals procedure; this is dealt with by the planning inspectorate (a separate government body).

If you are considering having building work carried out in and around your property, then you need to consider the following two main issues:

- Planning permission
- Building regulation approvals.

There are, of course, other considerations, as follows:

- The Party Wall etc. Act 1996
- Construction (Design and Management) Regulations 2007
- Listed buildings consent
- Rights of way.

The first two of these elements will be considered in detail later in this book (see pages 76–91).

Engage an architect, chartered surveyor or at the very least a competent person (see page 64) to assist with this stage. They will ensure a reasonably smooth path to achieving planning permission and gaining building regulation approvals prior to carrying out any work.

As with all building work, ultimately, you as the homeowner are responsible for complying with the relevant local authority's planning and building regulation permissions and approvals. Failure to comply with these rules and regulations could result in you being liable for any remedial action.

Discuss the building work proposals with your local authority's planning and building control departments prior to the commencement of any works.

BUILDER INFO

It is advisable to ask the client if they have discussed the proposals with their local authority before any work is undertaken. Put this request in writing and ask the client to respond in writing. If available, ask for the planning application or permit reference number and/or the building regulation approved document reference number for your files.

Relevant legislation

Planning

If you're considering work on your property, check if planning permission is required. Below are the relevant legislations that you may need to comply with:

- The Town and Country Planning Act 1990 (as amended by the Planning and Compulsory Purchase Act 2004)
- The Town and Country Planning (General Development Procedure) Order 1995 (to be amended by the Town and Country Planning (General Development Procedure) (Amendment) (England) Order 2008)
- Planning Act 2008.

Building

There are numerous pieces of legislation that could apply to buildings. The most important and relevant of these are listed below.

- The Building Act 1984 is the enabling Act under which the building regulations will have been prepared.
- The Building Regulations 2000 and Building (Approved Inspectors etc.) Regulations (amended) 2006, are set out under the Building Act 1984, and apply in England and Wales. They set standards for the design and construction of buildings to ensure the health and safety for people in or around those buildings. They also include requirements to ensure that fuel and power is conserved and that facilities are provided for people, including those with disabilities, to access and move around and inside buildings.
- Sustainable and Secure Buildings Act 2004.

The time period for obtaining planning permission is approximately six to eight weeks in most cases. The planning authority's decision can either permit or prevent a development. Building regulation approval exists in order to ensure that the works follow and comply with the relevant regulations. This approval can take three to five weeks, depending on the accuracy and comprehensiveness of the application and how busy your local authority building control department is at the time of your application.

Types of planning permission

Outline Planning Permission (OPP)

It is not always possible to present a proposed development as a whole, and initially only the principles of planning permissions are delivered. This will allow some components of the plan to be reviewed at a later date, depending on their importance to the overall plan. Issues such as access, routes, design and appearance may or may not be part of an outline planning application.

Upon granting of outline planning permission, a supplementary planning application will need to be made, which will incorporate previously reserved elements and allow planners to consider the plan in full. Development can only commence when this secondary application has been approved by the planners.

Outline planning status means that in principle it is possible to build on a particular piece of land and this status is usually valid for three years.

In some cases outline planning permission is necessary. However, in general, applying for full planning permission from the start is more time efficient and less costly.

Full Planning Permission (FPP)

In order to apply for full planning permission, a very detailed development plan will need to be drawn up that will leave no unanswered questions when it reaches the desk of the local authority planner. Works can commence as soon as full planning permission is granted.

It is normally advisable to apply for planning permission first, and either receive planning permission or be satisfied from discussions with the relevant planning officer that permission is on its way before applying for building regulation approval.

What is required in a full planning application?

Applications for full planning permission are required to be accompanied by the following:

- Application form (duly completed) to include:

    a) Applicant's details

    b) Agent's details

    c) Description of work

    d) Site address details.

- Location plan: Copies of a location plan based on an up-to-date map to a scale of 1:1250 or 1:2500 and showing at least two named roads and surrounding buildings. The application site should be clearly outlined in a red boundary.

- Site and other plans: Copies of a site plan should be submitted to a scale of 1:500 or 1:200, which should accurately show:

    a) The direction for north

    b) The proposed development in relation to the site boundaries and other existing buildings on the site, with written dimensions, including those to the boundary

    c) All buildings, roads and footpaths, including access arrangements

    d) All public rights of way

    e) The position of any trees on the land or adjacent to the buildings

    f) The extent and type of hard standing

    g) All boundaries, including walls and/or fencing.

- Existing and proposed floor plans and elevations to a scale of 1:50 and or 1:100. These drawings should clearly identify the existing property and the proposed work. They should also identify materials, windows and doors.

- Roof plans to the same scale to identify roofing materials and their position.

- Ownership certificates.

- Agriculture Holdings Certificate.

- The correct fee.

In some cases a Design and Access Statement must also accompany the application, unless it relates to one of the following:

- A material change of use of land and buildings
- Engineering or mining operations
- Householder developments.

Design and Access Statements are required for householder applications if the building falls within a designated area.

Applications for listed building consent will also require a Design and Access Statement. This should address the following issues:

- Special architectural interest
- Particular physical features of the building that justify its designation as a listed building
- The setting of the building.

The legislative requirements are set out in regulation 3A of the Planning (Listed Buildings and Conservation Areas) Regulations 1990. Three copies of all supporting documents must be submitted with each planning application. This is a statutory requirement for a valid application. However, some planning authorities may request additional copies.

In October 2008, new regulations came into effect – the Planning Act 2008. This was meant to streamline the planning process and allow various buildings and extensions and other works to become permitted developments, subject to limits and conditions.

> A Design and Access Statement is a short report accompanying and supporting a planning application. It should seek to explain and justify the proposals from a structural view. The Design and Access Statement should cover both the design principles and concepts that have been applied to the proposed development, as well as any issues relating to access to the development.

Some examples of permitted developments including limits and conditions are detailed below and on pages 63–71.

Local authority approvals for extension work

Under these new regulations that came into effect on 1 October 2008, an extension or an addition to a property is considered to be a permitted development, not requiring an application for planning permission, subject to the following limits and conditions:

- No more than half the area of land around the *original house* (see page 65) would be covered by additions or other buildings.
- No extension forwards of the principal elevation or side elevation fronting a highway
- No extension to be higher than the highest part of the roof
- Maximum depth of a single-storey rear extension of 3m for an attached house and 4m for a detached house
- Maximum eaves height of a single-storey rear extension of 4m
- Maximum depth of rear extension of more than one storey of 3m including ground floor
- Maximum eaves height of an extension 3m within 2m of boundary
- Maximum eaves and ridge height of extension no higher than existing house
- Side extension to be single-storey with maximum height of 4m and width of no more than half that of the original house
- Two-storey extension no closer than 7m to rear boundary
- Roof pitch of extension higher than one storey to match existing house
- Materials to be similar in appearance to existing property
- No verandas, balconies or raised platforms
- Upper floor side-facing windows to be obscured glass and any opening to be 1.7m above floor.

On *designated land* (see page 65) the following will apply:

- No permitted development for rear extension of more than one storey
- No cladding of the exterior
- No side extensions.

Building regulations

Most extensions will require approval under building regulations.

Local authority approvals for loft conversions

Planning permission is not normally required. However, permission is required where the roof space is to be extended or altered and the proposal exceeds specified limits and conditions.

Under the new regulations that came into effect from 1 October 2008, a loft conversion for a property is considered a permitted development, not requiring planning permission, subject to the following limits and conditions:

- A volume allowance of $40m^3$ for terraced houses
- A volume allowance of $50m^3$ for detached and semi-detached property
- No extension beyond the plane of the existing roof slope of the principal elevation that fronts the highway
- No extension to be higher than the highest part of the roof
- Materials to be of similar appearance to the existing house
- No verandas, balconies or raised platform
- Side-facing windows to be obscure glazed and any opening to be 1.7m above the floor
- Roof extension not to be permitted development in designated land (see page 65).
- Roof extension, apart from hip to gable ones, to be set back as far as is practicable, at least 20cm from the eaves.

Building regulations

Building regulations approval is required to convert a loft or attic into a living space.

Work on a loft, attic or roof may affect any one of several protected species but in particular may affect bats. Protected species must be considered before undertaking such work. A survey may be needed and, if bats are present, a licence may be required. Engage a competent person (see below) to assist with this stage. They will ensure a reasonably smooth path to achieving planning permission and gaining building regulation approvals prior to carrying out any building work.

Local authority approvals for electrical work

Planning permission is not generally required for installing or replacing electrical circuits. However, if works are being considered in a listed building, contact should be made with the local authority Planning Department before carrying out any work.

Building regulations

Any electrical works being undertaken in a home or garden in England and Wales have to comply with new rules in the building regulations and in particular part 'P'.

An installer who is registered with a Competent Person Scheme (BSI, CERTASS or FENSA) – a scheme that allows a variety of specified type of building work to be self-certified – should be employed to seek approval from the local authority building control department.

Local authority approvals for outbuildings

Rules governing outbuildings apply to sheds, greenhouses and garages as well as other ancillary garden buildings such as swimming pools, ponds, sauna or jacuzzi cabins, kennels, enclosures such as tennis courts and many other kinds of structures, the purpose of which is incidental to the enjoyment of the dwelling house. Other rules relate to the installation of a satellite dish, and the erection or provision of fuel storage tanks.

Under the new regulations that came into effect on 1 October 2008, outbuildings are considered to be permitted development, not needing planning permission, subject to the following limits and conditions:

- No outbuilding to be erected on land forwards of a wall forming the principal elevation
- Outbuildings and garages to be single storey with maximum eaves height of 2.5m and maximum overall height of 4m with a dual-pitched roof or 3m for any other roof
- Maximum height of 2.5m within 2m of a boundary
- No verandas, balconies or raised platforms
- No more than half the area of land around the *original house* would be covered by additions or other buildings
- In National Parks, the Norfolk Broads, Areas of Outstanding Natural Beauty and World Heritage sites the maximum area to be covered by buildings, enclosures, containers and pools more than 20m from the house to be limited to $10m^2$
- On designated land buildings, enclosures, containers and pools at the side of the property will require planning permission
- Within the curtilage of listed buildings, any outbuildings will require planning permission.

Building regulations

If the requirement is for a small, detached building such as a garden shed or summerhouse in the garden, building regulations will not normally apply if the floor area is less than $15m^2$.

If the floor area is between 15 and $30m^2$, building regulation approval will not normally be required provided that the building is either located at least 1m from any boundary or is constructed of substantially non-combustible materials.

In both cases, building regulations do not apply provided the proposed building does not contain any sleeping accommodation.

> The term *original house* means the house as it was first built or as it stood on 1 July 1948 (if it was built before that date). *Designated land* includes National Parks and the Norfolk Broads, Areas of Outstanding Natural Beauty, Conservation Areas and World Heritage sites.

Local authority approvals covering conservatories

Under these new regulations adding a conservatory to a residence is considered to be a permitted development, subject to the following limits and conditions:

- No more than half the area of land around the original house would be covered by additions or other buildings
- No extension forwards of the principal elevation or side elevation fronting the highway
- Maximum height of a single-storey extension of 3m for an attached house and 4m for a detached house
- Maximum depth of 3m for rear extension of more than one storey including ground floor
- Maximum eaves height of 3m within 2m of boundary
- Maximum ridge height no higher than existing house
- Side extension to be single storey with maximum height of 4m and width no more than half that of the original house
- Roof pitch of extension to match that of the original house
- No verandas, balconies or raised platforms.

Building regulations

Building regulations will normally apply to extensions; however, conservatories are normally exempt if the following rules apply:

- They are built at ground level and are less than 30m^2 in floor area
- At least half the new wall and three-quarters of the roof is either glazed or translucent material
- The conservatory is separated from the house by external quality door(s)
- Glazing and any fixed electrical installation comply with the relevant building regulations requirements.

Consideration should be given to problems associated with constructing a conservatory under first-floor windows. This may lead to a restricted ladder access or restriction of an escape route in the event of a fire.

Any new structural opening between the conservatory and the existing house will require building regulation approval.

Local authority approvals covering fences, gates and garden walls

Planning permission is required to erect or add to a fence, wall or gate if:

- It is over 1m high and next to a highway used by vehicles (or the footpath of such a highway) or 2m high elsewhere
- The property has an Article 4 direction applied (see below)
- The property is a listed building or in the curtilage of a listed building
- The fence, wall or gate of any other boundary involved forms a boundary with a neighbouring listed building or its curtilage.

> An article 4 direction prevents permitted development in designated land including national parks, Norfolk Broads, Areas of Outstanding Natural Beauty, conservation areas and World Heritage sites.

Building regulations

Fences, walls and gates do not require building regulation approvals. However, they must be structurally sound and maintained.

If the garden wall is classed as a 'party fence wall' and dependent on the type of work being considered, the neighbour needs to be notified under the Party Wall etc. Act 1996. (Wooden fences are not included.) See page 78.

Local authority approvals for kitchen and bathrooms

Planning permission for installing a kitchen or bathroom is usually not required unless it is part of a property extension.

If the property is listed, the local planning authority should be consulted.

Building regulations

A replacement kitchen or bathroom is unlikely to require building regulation approval. However, any new drainage or electrical work may require approval under the building regulations.

If a kitchen or bathroom is to be installed for the first time into an existing room then building regulations approval is likely to be required to ensure that the room is properly ventilated and the drainage and electrics meet the necessary regulations, along with structural stability and fire safety.

Local authority approvals for doors and windows

Planning permission is not usually required for repairing or replacing doors and windows. However, if the building is listed or in a conservation area, the local planning authority should be consulted.

Building regulations

Since April 2002, building regulations have applied to all replacement glazing. These regulations apply to thermal performance, safety, air supply, means of escape and ventilation.

An external door and window is a *controlled fitting* and as a result of this classification these regulations set out certain standards to be met when replacement is required.

It is permissible under these regulations to use a registered installer with a Competent Person Scheme (BSI, CERTASS or FENSA). In this case local authority building control permission will not be required.

> If a person who is not registered replaces any external door or window, they will need to apply for building regulation approval. On completion, a certificate of compliance should be obtained from the local authority building control department.

Thermal heat loss

Dwellings are required to be energy efficient. Steps are to be taken to reduce heat loss through glazing.

External doors and windows must comply with building regulation requirements. These regulations set a requirement for the amount of heat loss through glazing and framework: this is measured as a U-value.

Safety glazing

This is required as follows:

- Any glazing below 800mm from floor level.
- Any glazing within a window that is 300mm or less from a door and up to 1500mm from floor level.
- Within any glazed door up to 1500mm from floor level.

Ventilation
Windows and doors provide ventilation to rooms. Ventilation will be required to rooms, and certain rooms will require more ventilation than others. For example, kitchens and bathrooms require higher levels of ventilation (usually mechanical fans and opening windows). Trickle ventilation and/or opening windows are often adequate for other rooms.

Fire safety
Two aspects need to be considered:

- Fire spread between properties through unprotected areas.
- Means of escape in case of fire.

External doors and windows may need to have fire resistance and (in the case of doors) be self-closing. Some windows will need to be fixed closed.

Means of escape
When replacing windows, the opening should be sized to provide the same potential for escape as the window it replaces.

Means of escape should be considered for any new window to be installed in an extension or existing dwelling. It is good practice to replace any first-floor window with an escape window.

Criteria for means of escape windows:

- Width and height: either of these not to be less than 450mm.
- Clear opening area: not less than $0.33m^2$.
- Sill height: not less than 800mm and no more than 1100mm from floor level.

Access to buildings
When replacing any main entrance door in a dwelling that has been constructed since 1999, the threshold must be level, otherwise it will not comply with building regulations. This is to enable wheelchair access.

Local authority approvals for wind turbines

This subject is complex and evolving. It would be wise to assume that at present all wind turbines require planning permission.

It will be up to each local authority to decide what information is required to provide with an application. It would be wise to contact the local authority planning department to discuss all relevant issues, some of which may be as follows:

- Visual impact
- Noise
- Vibration
- Electrical interference
- Safety.

Building regulations

Building regulations will normally apply. Size, weight and force exerted on fixed points should be considered. Electrical installation will also need to comply with the relevant building regulations.

Local authority approvals for trees and hedges

A tree preservation order (TPO) is a form of planning control to protect trees that make an important contribution to their local surroundings. TPOs are made when trees are under threat of being cut down or damaged. It is illegal to cut down, prune or damage a tree protected by a TPO without consent from the local authority.

Trees overhanging boundaries

It is permissible to remove tree branches that overhang a boundary. The neighbour's permission is not required. However, it is always advisable to discuss the concerns with any neighbour.

High hedges

A complaint can be made to the local authority if a property is negatively affected by a neighbour's high hedge. A high hedge has to be at least 2m high and be mainly evergreen or semi-evergreen. Complaints will be dealt with by the local authority tree officers.

Local authority approvals for designated land

Additional forms of local authority approvals may be required for designated land such as the following:

- National parks, e.g. the Norfolk Broads
- Areas of Outstanding Natural Beauty
- Conservation areas
- World Heritage sites.

Local authority approval for listed buildings

If a property is listed grade one or two, the owner is required to obtain listed building consent (in addition to planning permission and building regulation approvals) from the local authority.

Failure to obtain or comply with planning permission

Failure to obtain planning permission or comply with the details of a planning permission is known as a 'planning breach'.

This usually occurs when:

- A development that requires planning permission is undertaken without permission being granted
- A development has been given permission subject to conditions and one of these conditions is broken.

> A planning breach itself is not illegal; the council will often permit a retrospective application where planning permission has not been sought. However, if the breach involves a previously rejected development (or the retrospective application fails) the council can issue an enforcement notice requiring the developer/homeowner to put things back as they were. It is illegal to disobey an enforcement notice unless it is successfully appealed against.

Applying for building regulation

There are two types of building regulation application forms:

Full plans submission

This type of application can be used for any type or size of work. Two copies of all plans and the appropriate plan fee will need to be submitted together with a completed application form. Four copies of plans are required for designated-use premises as covered by the Regulatory Reform (Fire Safety) Order 2005 (RRO). These are non-domestic premises such as:

- Hotels or boarding houses
- Factories
- Offices
- Shops
- Railway premises.

Building notice

This is an alternative method of application that does not require the submission of detailed plans. A dimensional block plan indicating position of site boundaries and drainage together with a building notice fee will be required to complete application. A building notice is best suited to minor residential work such as home alterations and small extensions and cannot be used for work involving designated-use premises or commercial use.

On achieving approval under one of the above submissions, work can commence. Notice must be given to the local authority building control department, as follows:

- Two days before commencement of works
- One day before concreting foundations
- One day before covering up foundation concrete
- One day before covering up any damp-proof course
- One day before over-site concrete is laid
- One day before covering up drainage
- No more than five days after covering up drains (drain test)
- One day before covering any structural timber, steelwork or concrete work
- Not less than five days before occupation
- No more than five days after completion of the work.

Remember that it is your responsibility to ensure that building control has been notified. Even if you have asked the builder or tradesman to do this, it is your responsibility to check that it has been carried out.

The main function of your local authority building control is to ensure compliance with the building regulations.

In complying with the building regulations as set out in the Building Act 1984, there are 18 approved documents to consider and comply with, as described in the box below and on pages 74–75. The building regulations are divided into five schedules. Schedule 1 is the key schedule for carrying out building work.

Schedule 1 – Requirements

A. *Structure*:

- This supports Schedule 1, A1, A2 and A3:
- Section 1 gives details of codes and standards that can be used for building standards and basic principles.
- Section 2, which deals with houses and other small buildings, contains guidance on sizing of timber members, wall thicknesses, masonry, chimneys and concrete foundations.
- Sections 3 and 4 covers wall cladding and roof covering.
- Section 5 deals with disproportionate collapse.

B. *Fire safety:*
Approved document B has been divided into two volumes: volume 1 deals with the dwelling house; volume 2 deals with all other buildings. This is the most complex of the regulations, but suffice to say it deals with fire safety.

A new regulation 16B has been introduced to make sure sufficient information is recorded to assist the eventual owner/occupier/employer to meet their statutory duties under the Regulatory Reform (Fire Safety) Order 2005.

C. *Site preparation and resistance to contamination and moisture:*

- Section 1 deals with clearance or treatment of unsuitable materials.
- Section 2 deals with contamination.

- Section 3 deals with subsoil drainage.
- Sections 4 to 6 deal with and explain the measures necessary to prevent moisture ingress, including condensation.

D. *Toxic substances:* This supports Schedule 1, Part D. This schedule gives advice on guarding against fumes from urea formaldehyde foam.

E. *Resistance to the passage of sound:* This supports Schedule 1, Part E and deals with the building's ability to prevent the passage of unwanted sound from internal sources. It must be noted that sound penetration through external walls is covered by the planning legislation and not the building regulations.

F. *Ventilation:* Supports Part F Schedule 1, this covers means of ventilation and applies to all building types, including additions and extensions.

G. *Hygiene:* Supporting Part G of Schedule 1, this deals with water closets and bathrooms, as well as unvented hot water systems.

H. *Drainage and waste disposal:* This supports Part H of Schedule 1 and covers above as well as below ground drainage, as well as waste-water treatment systems and cesspools.

J *Combustion appliances and fuel storage systems:* Supporting Part J of Schedule 1, this covers gas appliances up to 60kW and solid and oil fuel appliances up to 45kW, as well as protection of liquid fuel storage and pollution caused by heating oil leakage.

K. Protection from falling, collision and impact. Part K of this schedule 1 covers design and construction of stairs, ramps and guarding.

L. *Conservation of fuel and power:* Supporting Part L, Approved Document L is divided into four separate documents dealing with conservation of fuel and power in new and existing dwellings. Making sure they are reasonably efficient in their use of energy and that carbon dioxide emissions are kept to a minimum.

M. *Access to and use of buildings:* Supporting Part M of schedule 1, this gives practical assistance on means of access to the building as well as passenger lifts and common stairs.

N. *Glazing – safety in relation to impact, opening and cleaning:* Supporting Part N, this covers safe operation and access for cleaning windows, as well as measures to reduce risk of accidents.

P. *Electrical safety:* Supporting Part P, this approved document gives guidance on design, installation and testing of electrical installations.

There are three further approved documents:

1. Materials and Workmanship to support Regulation 7

2. *Timber Intermediate Floors for Dwellings*, published by Timber Research and Development Association

3. *Basements for Dwellings*, published by the British Cement Association

Apart from dealing generally with the standard of materials and workmanship needed with building work, Materials and Workmanship to support Regulation 7 is also concerned with the following:

- The use of materials that are susceptible to change
- Resistance to moisture and harmful substances in the sub soil
- Materials with a short life-span.

The use of materials that are unsuitable for permanent buildings is covered by section 19 of the Building Act 1984. Local authorities are enabled to reject plans for the construction of buildings using materials of a short life span or otherwise unsuitable materials.

Conclusion

Gaining local authority permissions and approvals can be a complex stage of any building project and can be absolutely crucial to getting things right for the start of the works stage.

Engage an architect, chartered surveyor or at least a competent person to assist with this stage. They will ensure a reasonably smooth path to gaining local authority approvals.

Party Wall etc. Act 1996

This Act provides a way forwards for preventing and resolving disputes in connection with party walls, boundary walls and excavations near neighbouring properties. The Act of 1996 was based on the London Buildings Act, which applied to inner London for many decades. (This Act replaced those provisions in inner London.)

Any person proposing to carry out work anywhere in England and Wales that may affect their neighbouring properties, as described in the above Act, must give owners notice of their intentions. A notice must be given even where the work will not extend beyond the centre line of a party wall.

Adjoining owners can agree or disagree with what is proposed. Where there is a disagreement, the Act provides for the resolution of disputes.

The Party Wall etc. Act 1996 covers:

- Various works that are going to directly affect an existing party wall
- New building or extensions at or astride the boundary line between properties
- Excavations within 3m or 6m of a neighbouring building or structure. This will depend on the depth of the hole or foundations.

This chapter is primarily aimed at your client, since it is they who should negotiate with their neighbours over party walls etc. prior to building work commencing. However, much of the information herein is key to the overall building process and should be duly noted and acted upon by you.

What is a party wall?

This Act recognizes two main types of party wall. These are referred to as a 'party wall' or a 'party fence wall'.

Party wall

A wall is a 'party wall' if it:

a) Forms part of a building and stands astride the boundary of land belonging to two or more different owners (see diagram 1).

b) Separates buildings and stands astride the boundary of land belonging to two or more different owners (see diagram 2).

Always check with the client that they have given notice to their neighbour, as required by the Party Wall etc. Act 1996.

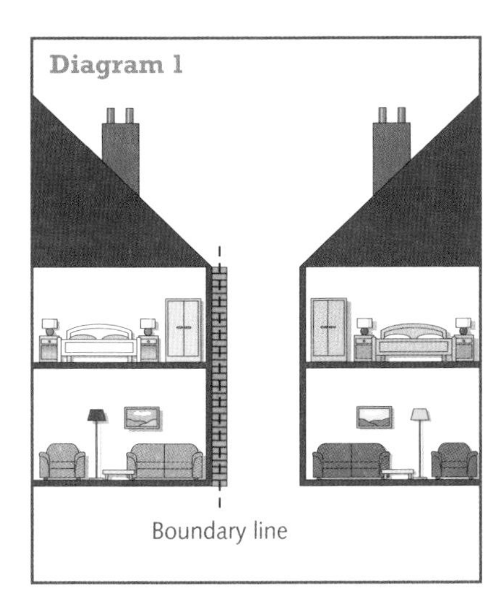

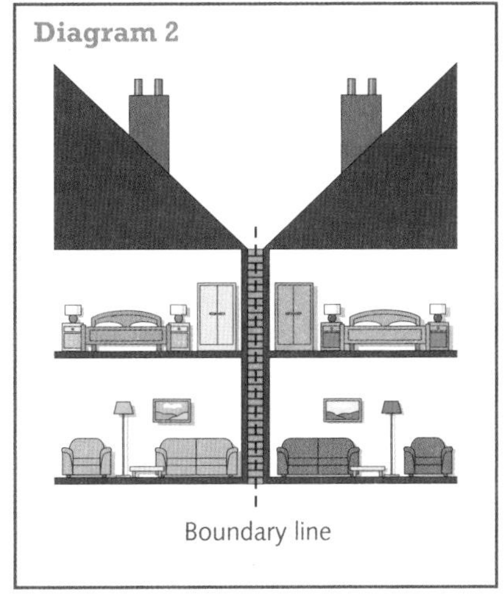

c) Stands entirely on one owner's land but is used by two or more owners to separate their buildings. Where one person has built the wall in the first place and another has butted their building up against it without constructing their own wall, only the part of the wall that does the separating is 'party', i.e. the sections on either side or above are not 'party' (see diagram 3).

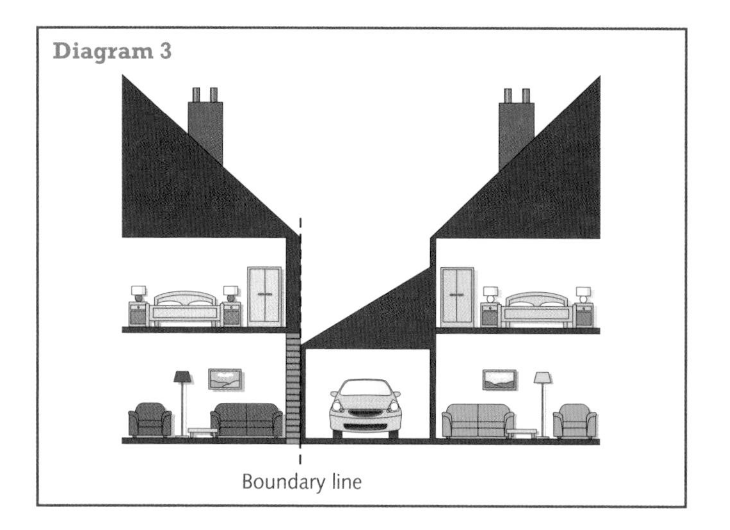

Party fence wall

A wall is a 'party fence wall' if it is a wall that is not part of a building, that stands astride the boundary line between grounds of different owners and is used to separate those grounds; for example, a garden wall (see diagram 4). This does not include such things as wooden fences.

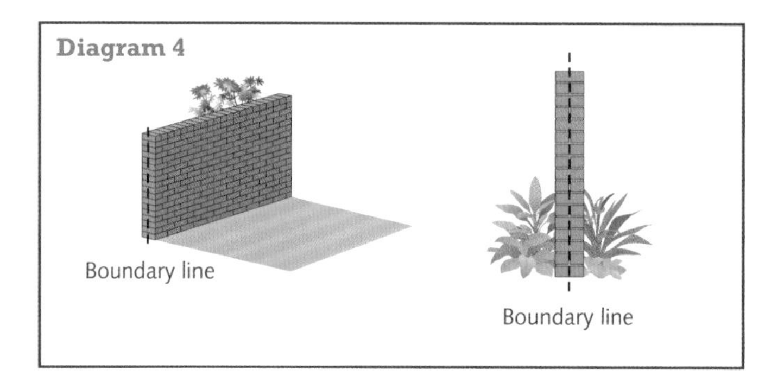

The Act also makes mention of a 'party structure'. This is a wider term that could be a party wall or a floor partition or other structure that separates a building or parts of a building approached by a separate staircase or entrance; for example flats (see diagram 5).

Diagram 5

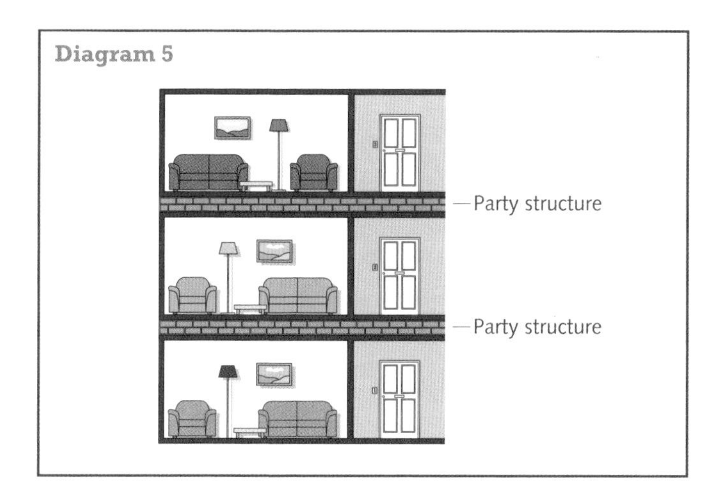

Work on an existing party wall (Section 2 of the Act)

The Act provides a building owner who wishes to carry out various sorts of work to an existing party wall with additional rights to do so. These go beyond ordinary common law rights.

Section 2 of the Act lists work that can be done. The most commonly used rights are:

- To cut into a wall to take the bearing of a beam (RSJ or universal beams), or to cut into a wall to insert a damp-proof course
- To raise the whole party wall and, if required, to cut off any projections that prevent you from doing so
- To demolish and rebuild the party wall
- To underpin the whole party wall
- To protect two adjoining walls by putting a flashing from the higher over the lower.

Minor works on a party wall

Minor works on a party wall may be considered to be too trivial to come under the Act. For example:

- Drilling into your own half of a party wall to fix plugs and screws for standard wall units or shelves
- Drilling into your own half of a party wall to add/replace recessed electrical wiring or sockets
- Re-plastering.

The key point to consider is whether your planned work might have consequences for the structural strength and support functions of the party wall. If in doubt, seek advice from a qualified building professional.

It would also be advisable to let your neighbours know that you are going to carry out some work.

Always treat the neighbours with respect. If working on or close to the boundary, ensure the neighbour's side is kept clean and tidy. On completion, carry out any additional work required to return the neighbour's side to its original condition.

Duties under the Act

If you are planning work covered by Section 2 of the Act, you must inform all adjoining owners. It is best to discuss your plans fully with your neighbour before you or your professional adviser give notice in writing. If you have already ironed out any concerns your neighbour may have, this should give them enough confidence to give their written consent readily. You do not need to engage a professional adviser.

Your written notice must include the following details:

- Your own name and address
- The building's address
- A clear statement that your notice is a notice under the provision of the Act
- Full details of what you propose to do. Include any plans, if available
- When you propose to start the works.

You may deliver the notice in person or post it. You do not need to inform the local authority. If the property is empty or the owner is not known you may address the notice to 'the owner' of the premises and fix it to an easily visible part of the premises.

You need to give your adjoining owner at least two months' notice before starting any of the work. The notice is only valid for a year.

After serving the notice

A person who receives the notice about the intended work may give their consent in writing, or give a counter-notice, setting out what additional or modified work they would like to be carried out. This counter-notice should be received within 14 days.

If after 14 days no consent or notice has been received, it should be assumed a dispute has arisen.

However, if you do receive a counter-notice within 14 days, you must respond further within 14 days. Again, if you do not respond within this period it will be assumed that both parties are in dispute.

The best way to sort out any differences of opinion is to have a friendly chat. Then any agreement between the parties should be confirmed in writing. Always respect your neighbour's concerns.

If agreement cannot be reached

If an agreement cannot be reached with your neighbour, the next best thing is to appoint, jointly, what the Act calls an 'agreed surveyor' to draw up an award. The 'agreed surveyor' should not be the same person employed to supervise the work.

Alternatively, each neighbour can engage their own surveyor to draw up the award together. The two surveyors will nominate a third surveyor who would be called in only if the two surveyors cannot agree.

What is included in the surveyor's award?

The party wall award is a document that includes the following:

- Sets out the work to be executed
- Sets out when and how the work is to be carried out (working hours etc.)
- Records the conditions of the adjoining property before works commence and again on completion (so that any additional damage can be attributed and made good by the owner who is having the work carried out)
- Allows access to the surveyors to inspect the works while they are ongoing.

Who pays the surveyor's fees?

It is usually the owner who is having the work carried out who should pay all of the costs associated with drawing up the award.

Is the surveyor's award final?

Either side has 14 days to appeal to the county court against the award.

Who pays for the building work?

The award will set this out. In general, the Act says that the building owner who is having the work carried out should pay. There are cases when the adjoining owner should pay, for example:

- Where work to the party wall is needed because of maintenance reasons
- Where the adjoining owner requests that additional work should be done.

The award may deal with the apportionment of cost of the work. The dispute procedure may be used specifically to resolve the question of cost.

What if the neighbours will not cooperate?

If the adjoining owner refuses to discuss your proposals and refuses to appoint a surveyor, the person having the work carried out can appoint a second surveyor on the adjoining owner's behalf – this will allow the procedure to go ahead.

Can the adjoining owner refuse access?

Under the Act, an adjoining occupier must, when necessary, let the surveyors and workmen onto his premises after the 14 days notice has been given.

> It is an offence, which can be prosecuted in the magistrates' court, to refuse entry to or obstruct someone who is entitled to enter the premises under the Act.

If the adjoining property is empty, the workmen and the surveyors may enter the premises if they are accompanied by a police officer.

Building on the boundary line between neighbouring pieces of land (Section 1 of the Act)

If it is planned to build a party fence wall astride the boundary line, you must give notice as explained previously with the exception that at least one month's notice must be given.

There is also no right to build astride the boundary if your neighbour objects. You must also give notice if it is planned to build a wall on your own land but up against the boundary line.

The Act contains no enforcement procedure for failure to serve a notice. However, if the work is started without first giving notice in the proper way, adjoining owners may seek to stop the work through a court injunction or seek legal redress.

Work can commence one month after the notice has been served. This work may include foundations that extend under the adjoining owner's land.

The wall will be built wholly at the owner's expense and the owner will be expected to compensate any adjoining owner for damage to the adjoining property caused by the building work.

Excavation near a neighbouring building (Section 6 of the Act)

The adjoining owners must be served a notice (as previous paragraphs) and the notice must state whether the foundations are to be strengthened or protected if one of the following apply:

- It is planned to excavate or construct foundations for a new building or an extension within 3m of a neighbouring owner's building or structure where the work will go deeper than the neighbour's foundations (see diagram 6).

Care should be taken when excavating close to a neighbour's property, as vibration can cause internal and external damage. Carry out close inspection for existing damage prior to commencement. If a surveyor is engaged under the Party Wall etc. Act 1996 they will carry out a condition survey that will cover this aspect.

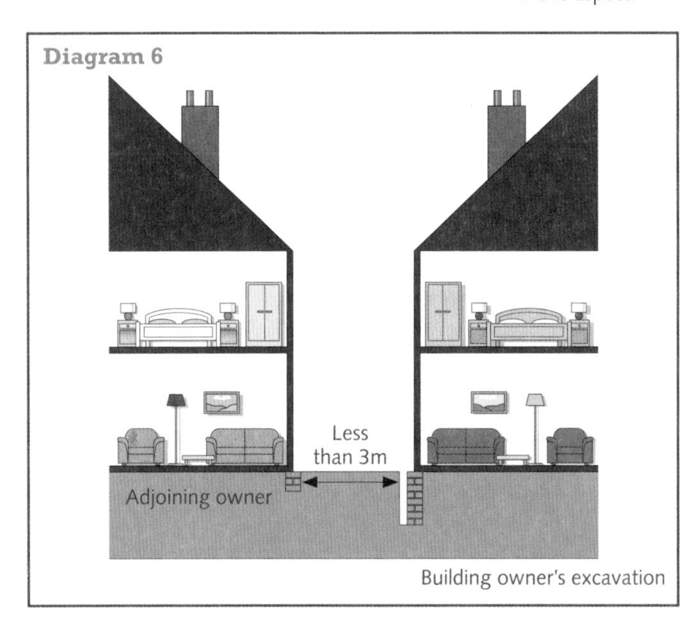

• It is planned to excavate or construct foundations for a new building or an extension within 6m of a neighbouring owner's building or structure where the work will cut a line drawn downwards at a 45-degree angle from the bottom of the neighbour's foundation (see diagram 7).

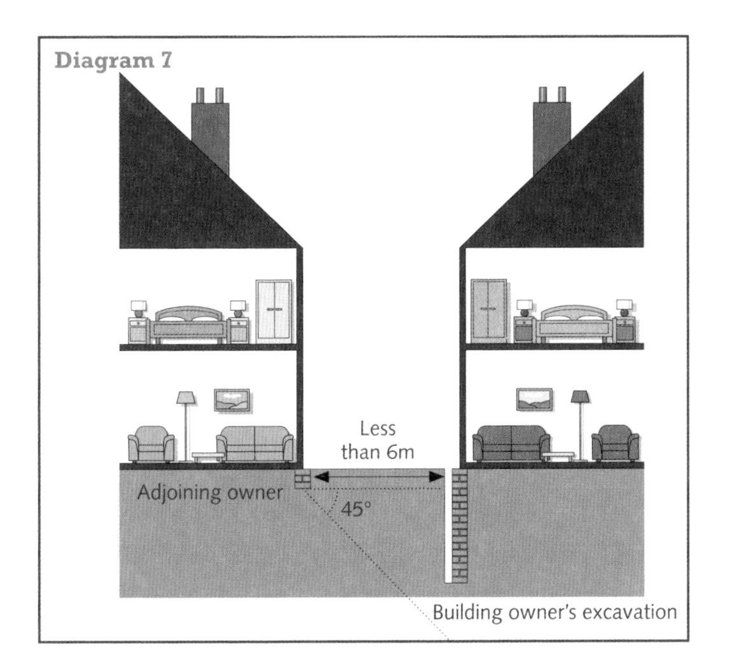

The Act contains no enforcement procedure for failure to serve a notice. However, if the work is started without first giving notice in the proper way, adjoining owners may seek to stop the work through a court injunction or seek legal redress.

Again, at least one month's notice must be given before any work may commence.

It should always be treated as a priority to discuss any work you are having done on your property with the adjoining neighbours to alleviate any concerns they may have. This can save time and problems later.

The Construction (Design and Management) Regulations 2007

The above regulations (CDM 2007) came into force on 6 April 2007. These regulations replaced the previous regulations CDM 1994 and the Construction (Health, Safety and Welfare) Regulations 1996. These new regulations are supplemented by the New Approved Code of Practice entitled Managing Health and Safety in Construction.

The regulations strengthen the duties of all involved in the design and construction of projects of every size. There is a greater obligation on clients and employers who can no longer assign their duties to agents. If duties are not met, they can be summoned under criminal and civil liability. The CDM 2007 Regulations are divided into five parts:

- Part 1 deals with the application of the regulations and definition
- Part 2 covers general duties that apply to all construction projects
- Part 3 contains additional duties that only apply to notifiable construction projects, i.e. those lasting more than 30 days or involving more than 500 person days of construction work
- Part 4 contains practical requirements that apply to all construction sites
- Part 5 contains the transitional arrangements and revocations.

CDM 2007 Regulations do not apply to the private homeowner unless they act as a developer, i.e. carrying out work as a business or for profit.

Although the CDM 2007 Regulations are primarily aimed at builders and tradesmen, if you are a developer it is worth familiarizing yourself with them if you are in any doubt about the way in which your builder/tradesman is undertaking their responsibilities. Specific points of interest are brought to your attention later in this chapter.

Construction remains a disproportionately dangerous industry in which improvements in health and safety are urgently needed. The improvements require significant and permanent changes in duty holder (those with legal duties) attitudes and behaviour. Since the original CDM Regulations were introduced in 1994, concerns were raised that their complexity and bureaucratic approach for many duty holders frustrated the Health and Safety Executive (HSE) who polices these regulations. These views were supported by an industry-wide consultation in 2002, which resulted in the decision to revise the regulations.

The new CDM 2007 Regulations revise and bring together the CDM Regulations 1994 and the Construction (Health, Safety and Welfare) Regulations 1996 into a single regulatory package.

The main aims of these regulations

- To reduce bureaucracy and paperwork, allowing improvements in the planning and management of health and safety in projects from the very start.
- To simplify the regulations in order to improve clarity, making it easier for duty holders to know what is expected of them.
- To encourage more integration. Strengthening the requirements regarding co-ordination and cooperation, particularly between designers and contractors.
- To simplify the assessment of competence (both organizations and individuals) in order to help raise standards and reduce bureaucracy.
- To ensure that the right people are in place for the right job at the right time in order to manage and identify risks early on so that these can be reduced or eliminated. Any remaining risk should be properly managed.
- To maximize their flexibility to fit with the vast range of contractual arrangements.
- To target effort where it can do most good in terms of health and safety.
- To discourage unnecessary bureaucracy.

Everyone in construction needs to know about the new construction health and safety regulations.

Notification

Shortly after appointing the CDM co-ordinator, the duty holder should notify the local Health and Safety Executive (HSE) office that covers the site where the construction work is being carried out. This can be done on Form F10, available from the HSE. From June 2008 a new interactive F10 EForm is available on the HSE website for duty holders to complete and submit construction project notifications online.

Sites that do not require notification are as follows:

- When the project is for a domestic client
- When the project does not last more than 30 working days
- When the project does not involve more than 500 person days; say, 10 people working for less than 50 working days.

All other construction projects should be notifiable, but if in doubt contact the HSE for further advice.

If the work is carried out under CDM 2007 Regulation then it is up to you to give the architect/CDM co-ordinator at the outset as much relevant information as possible, i.e. services routes, drainage runs, any asbestos on the premises, etc.

Clauses relevant to developers are denoted by an arrow in the following list of information.

Duty holders under the CDM 2007

CDM 2007 places legal duties on virtually everyone involved in construction work. Those with legal duties are commonly known as duty holders, who under the CDM Regulations are as follows:

Clients

A client is anyone having construction or building work carried out as part of their business. This could be an individual, a partnership or a company and includes property developers or management companies for domestic properties.

CDM co-ordinators

A CDM co-ordinator has to be appointed to advise the client on projects that last more than 30 days or involve 500 person days of construction work. The CDM co-ordinator's role is to advise the client on health and safety issues during the design and planning phases of the construction work.

Designers

The term 'designer' has a broad meaning and relates to the function performed, rather than the profession or job title. Designers are those who, as part of their work, prepare design drawings, specifications, bills of quantities and specification of articles and substances. They could include architects, engineers and quantity surveyors.

Principal contractors

A principal contractor has to be appointed for projects that last more than 30 days or involve more than 500 person days of construction work. The principal contractor's role is to plan, manage and co-ordinate health and safety while construction work is being undertaken. The principal contractor is usually the main managing contractor for the work.

Contractors

A contractor is someone who runs a business that is involved in construction, alteration, maintenance or demolition work. This could involve building, civil engineering, mechanical, electrical, demolition and maintenance companies, partnership and the self-employed.

Workers

A worker is anyone who carries out work during the construction, alteration, maintenance or demolition of a building or structure. A worker could be, for example, a plumber, electrician, scaffolder, painter, decorator or steel erector, as well as those supervising the work, such as foremen and chargehands.

It is the duty of the client and indeed all the duty holders to ensure the competency and the resources of all appointed personnel. The client must make sure that sufficient time and resources are allowed for all stages of the work to be carried out safely.

The health and safety file

All notifiable projects should have a health and safety file containing the information needed to allow future construction work, including cleaning, maintenance, alteration, refurbishment and demolition to be carried out safely. It should alert workers to risk and help them to work safely. The file should prove useful to:

- Clients who have a duty to provide information about their premises to those who carry out such work there
- Designers during development of further designs or alterations
- CDM co-ordinators preparing for construction work
- Principal contractors and contractors preparing to carry out or manage such work. The file should form a key part of the information that the client, or the client's successor, is required to provide for future construction projects. The file, therefore, should be kept up to date after any relevant work or survey.

The scope, structure and format for the file should be agreed between the client and the CDM co-ordinator at the start of a project.

Clients, designers, principal contractors, other contractors and CDM co-ordinators all have legal duties in respect of the health and safety file.

- CDM co-ordinators must prepare, review, amend or add to the file as the project progresses and hand it over to the client on completion.
- Clients, designers, principal contractors and other contractors must supply the relevant information necessary for compiling or updating the file.
- Clients must keep the files to assist with future construction work.
- Everyone providing information should make sure that it is accurate and provided promptly.

If you are a developer it is advisable to read the information given on this page and the facing page thoroughly, as these important regulations affect everyone involved in a building project regardless of its scale or nature.

The contents of the health and safety file

Consider the following factors, if relevant to the work in hand:

- A brief description of the work to be carried out
- Any hazards that remain and how they have been dealt with
- Key structural principles, such as safe working loads for floors and roofs – particularly where these may preclude placing scaffold or heavy machinery
- Hazardous materials used and found, including, for example, lead paint, pesticides, special coatings that should be burnt off, etc.
- Information on dismantled or installed plant
- Health and safety information about equipment provided for cleaning or maintaining the structure
- The nature and location of markings of significant services, including underground cables, gas supply and fire-fighting services and equipment
- Information such as drawings of the structure as-built, its plant and equipment.

Storing the file after the work is complete

In order for the health and safety file to be useful, it needs to be kept up to date and retained for as long as it is relevant. This is normally the lifetime of the building. It may be kept electronically (with suitable back-up arrangements), on paper or film or in any other durable format. In cases in which clients dispose of the entire interest in the structure, they should pass on the file to the new owners and ensure that they are aware of the nature and purpose of the file. If they sell part of a structure, any relevant information in the file should be passed or copied to the new owner.

Conclusion

The above is a brief look at the Construction (Design and Management) Regulations 2007 and some of the important elements of the regulation. For more detailed information, read Managing Health and Safety in Construction, Construction (Design and Management) Regulations 2007 Approved Code of Practice, which is available from the Health and Safety Commission.

Insurances and Guarantees

In this chapter we examine insurances and guarantees for basic building work around the property. These take a variety of forms and serve a number of different purposes. It is the builder/tradesman's responsibility to ensure that they have all the cover they need.

Introduction to liability insurance

If someone is injured or their property is damaged, the person or business responsible may be sued and held legally liable for the injury or damage to the property.

Where legal liability is established, damages will be awarded to the claimant to compensate them for their injury or damage to their property.

Where damages for an injury have been paid, it should be noted that the NHS is entitled to recover the cost of hospital treatment, including ambulance costs. The person or business responsible will also need to pay legal costs, including those of the claimant's.

Liability insurance is designed to protect your business against these costs.

Public liability insurance

Public liability insurance protects you and your business against claims brought against you or your business by other people. It covers you for any injury caused to a third party or damage to their property while you are going about your business activities. It also covers any related legal fees, costs and expenses as well as the costs of any hospital treatment, including ambulance costs.

It is very important that your builder or tradesman is properly insured and that they can offer all necessary guarantees and warranties for their work. Although the contents of this chapter are primarily aimed at builders who want to ensure they are correctly covered and indemnified, as a homeowner it is in your interest to familiarize yourself with the different types of insurance and guarantees and then to ensure that your builder has all the appropriate certification for the job in hand.

The cover also extends to cover products supplied by you in your business. The level of cover is usually £1m, £2m or £5m.

Public liability insurance is not a legal requirement, but it makes good business practice to invest in it. Local authorities will generally demand a minimum level of £2m cover for works to be undertaken at their premises or on their behalf.

Employers' liability insurance

Employers' liability insurance enables businesses to meet any costs of damages and legal fees for employees who are injured or made ill at work through the fault of the employer. Employees injured due to an employer's negligence can seek compensation, even if the business goes into liquidation or receivership. The NHS can also claim costs of hospital treatment (including ambulance costs) when personal injury compensation is paid. This applies to incidents that occurred on or after 29 January 2007.

It is a legal requirement to hold employers' liability insurance following legislation passed by the government in 1969. The following rules dictate whether employers' liability is required.

- If you are not a limited company then employers' liability insurance is required whenever you have staff – that is, people who work for you, excluding business partners or family members. Labour-only sub-contractors are considered to be staff. These include any person working under your guidance and using your tools and equipment.
- If you are a limited company consisting only of yourself, and manual and clerical staff, where you own more than 51 per cent of the business, then employers' liability insurance is not required.
- If you are a limited company consisting only of yourself, and manual and clerical staff, but own less than 51 per cent of the business, then employers' liability insurance is required.

> Advice on employer's liability insurance and the level of appropriate cover required by the law should be sought from an 'authorized insurer' working under the terms of the Financial Services and Markets Act 2000. The Financial Services Authority (FSA) maintains a register of authorized insurers. Check details of the FSA's website at www.fsa.gov.uk or telephone the FSA on 0845 6061234.

- If you are a limited company with more than one employee, then employers' liability insurance is always required.
- As a rule of thumb, if you deduct national insurance and income tax from your employee(s), then you need employers' liability.

The legal level of cover is £5 million. However, a cover of £10 million is advisable.

> The Health and Safety Executive (HSE) is responsible for enforcing the law on employer's liability insurance. You or your business can be fined up to £2,500 for each day that you do not have the appropriate insurance.

Professional indemnity insurance

If you are in the business of selling your knowledge or skills, you may want to consider taking out professional indemnity insurance.

This type of insurance protects your business against claims for loss or damage by a client or third party if you have made mistakes or are found to have been negligent in some or all of the services that you provide. This insurance will also cover legal costs.

It is advisable to make sure you are properly covered, as this type of insurance cover is usually on a 'claims made' basis. This means that the policy will only respond to claims that are made while the policy is live. If you change insurers, you will need to arrange for your new insurers to accept new claims for prior incidents.

It is also advisable to keep all projects well documented. Ensure that you set out specific responsibilities in your contracts with clients beforehand and deal with complaints promptly.

You should always notify your insurance company before commencing any building work on your property. Most building insurance policies come with third-party cover, just in case an injury is caused to a neighbour by existing material falling from your property. By notifying the insurance company they have the option of considering the risk, and whether it has increased or not by having building work carried out. If you do not notify your insurance company, your insurance cover may be invalidated.

Directors' and officers' liability

Directors and officers of a business have various duties, responsibilities and powers in connection with their position. In most cases these are set out in a job description or terms of reference. As a result, they can be held responsible for a range of issues such as:

- Health and safety
- Data protection
- Maintaining satisfactory accounts
- Fraud
- Negligence.

If your directors or officers have inadvertently acted outside their terms of reference and this gives rise to a claim, then compensation and legal fees will be covered by this insurance. If, however, the act was deliberate, then it may not be covered by the policy.

Contractors all risk insurance

In addition to the above insurances, there is an insurance cover called 'contractors all risk'. This type of policy can vary but generally covers such elements as follows:

- Design work
- Defects in materials or workmanship
- Theft of materials and plant.

The advantage of this type of policy is that if renewed annually it will cover work carried out in the past, so that if a claim is made some time after the work has been completed, this insurance will cover the claim.

In practice, this is the only type of insurance that will cover any latent defect. If dealing directly with an insurance company it would be wise to check that they are a member of the Association of British Insurers (ABI).

> It is always wise to discuss the nature of your business with an insurance broker who can then identify your business insurance needs. It is important to have peace of mind, allowing you to concentrate on the work in hand.

Guarantees and warranties

A guarantee:

- Is usually offered free of charge
- Is a promise from the tradesman to sort out any defects within a fixed period of time
- Is a legally binding contract even if you did not pay for it
- Must explain how to make a claim in a way that can be understood
- Should add to, not take away from, your rights under consumer law
- Works whether or not you have a warranty.

A warranty:

- Is like an insurance policy for which you must pay a premium
- Is sometimes called an extended guarantee
- Might cover a longer period than a guarantee and might cover a wider range of problems
- Is a legal contract, so you can take the company to court if they do not honour it
- Should contain terms that are clear and fair
- Does not diminish your rights under consumer law
- Can run alongside a guarantee.

Trade association guarantees or warranties

Some trade associations offer as a benefit to their members the option of an insurance-backed guarantee or warranty. They may also offer a mediation or conciliation service to assist should a dispute arise. An insurance-backed guarantee or warranty is usually set up for a small percentage of the contract sum and covers the required work for a set period.

The information provided here on guarantees and warranties is as relevant to you as it is to the builder or tradesman so it is best to familiarise yourself with the details.

Some manufacturers of good-quality materials also offer guarantees and these are usually free of charge.

Some local authority building control departments may offer an insurance warranty scheme based on their inspection criteria, or they may have a scheme in which they accept a trade association's warranty scheme.

Trade association guarantee or warranty schemes can cover the following elements:

- Deposits paid to the builder
- Defects due to bad workmanship or poor-quality material (claims for this element are usually restricted to claims within the first year or two)
- Structural failure
- If a builder goes out of business an alternative builder can be appointed to complete the works.

Always make sure that the cover extends to sub-contractors' and client's supply items.

Any guarantee or warranty covers the property so check that, if the owner decides to sell and move on, the insurance policy can be passed on to the new owner. While these schemes charge a small percentage of the contract sum, they can offer peace of mind.

A written guarantee is only as good as the company that offers it. If a company goes out of business, it is unlikely the written guarantee will count for anything. It may, however, offer you short-term peace of mind. Always make sure it is offered in writing.

An insurance-backed guarantee or warranty scheme should be seriously considered if any work is being carried out around the property because of the many problems encountered (disputes and poor workmanship issues) within the building industry. If the company goes out of business it is likely you will be covered by the insurance company. Make sure a copy is received on commencement and handed over on completion. Always check with the insurance company that a guarantee or warranty is in place before the commencement of any works.

Standard Terms and Conditions

When carrying out minor work around a property for a client, whether private or commercial, it is recommended that terms and conditions of trading are attached to an estimate or quotation to cover all aspects of business trading. If the works are of a reasonable size, such as an extension and/or alterations, then a contract can be agreed and put in place.

Often, small works, such as internal plaster repairs to a room, or perhaps the easing and adjusting of a door or window, are carried out prior to redecorations by others. The value of such work may be somewhere between £75 and £500. This work can be carried out following an estimate or quotation or, if the works are required quickly and it is difficult to provide a fair and reasonable estimate or quotation, the works can be carried out on a time-and-material basis. However, this method needs careful planning so that you can arrive and get on with your work. You need to keep an accurate record of time taken and materials supplied and agreed with the client. You are, of course, entitled to a reasonable allowance for overheads and a reasonable profit margin.

In this instance, where the works are of a small nature, you should include standard terms and conditions of trading within the estimate or quotation.

As this is an industry featuring lots of different types and sizes of business, it is up to you to decide which terms and conditions to use. The following list is by no means exclusive, but gives some examples of relevant terms and conditions.

It is important to know exactly where you stand before any work commences on your property, so it is as well to read all terms and conditions attached to any estimate or quotation or whenever they form part of a formal contract between you and a builder.

Recommended terms and conditions

1. Payment clause

- Estimate/quotation, inclusive of VAT.
- Three options: Option A – payment of a lump sum on completion; Option B – payment by stage payments; and Option C – payment at periodic intervals, based on work carried out to date.
  Option [A/B/C] applies.
- A retention sum of five per cent of the agreed price should be paid within one month of completion of the agreed work, provided all snagging issues have been attended to.

2. Variations

- Any requested variation cost is to be agreed with the customer in writing before works are carried out.
- If the customer requests any variations, the agreed price will be varied by an amount or time period (as the case may be) to be agreed between the parties.
- Requested variations and their effect will be recorded.

3. Programme of work

- A commencement and completion date will be agreed before commencement.
- If warranted, key dates will be agreed throughout the programme of the agreed work. Delivery dates of key elements will be mentioned.

Specific clauses or points of particular relevance to you as the homeowner are denoted by an arrow in the following lists of information.

4. Workmanship and materials supplied

- The builder/tradesman shall carry out the agreed work in a proper and workmanlike manner.

- All materials supplied by the builder/tradesman will also be of satisfactory quality and suitable for their intended purpose. Materials must be new unless otherwise agreed with the customer.
- Any faulty workmanship and or materials must be notified to the builder/tradesman as early as possible. The builder/tradesman must be given the opportunity to inspect and, if deemed necessary, request an expert's report to qualify problems and faults. If found to be as recorded, then the builder/tradesman will repair or replace materials or work as required.

- Manufacturer's recommendations will be followed and the customer must adhere to any necessary recommendations.

- Any materials supplied by the customer must be delivered to the premises and checked by the customer prior to incorporating into the agreed work. Any delay caused by the customer's supply may be charged for.

5. Weather

- If the builder/tradesman cannot complete the agreed work on time because of extremely inclement weather, an extension to the agreed period will be agreed within a fair and reasonable time.
- No concrete work, brickwork or plastering and/or external works such as rendering, slab laying or decorations will be carried out if temperatures are 2°C or below.

6. Water, electricity, gas and/or oil

- The customer will supply all necessary water and electricity for the carrying out of the agreed work without charge to the builder/tradesman.

- Any movement of meters or mains from the boundary to the meter and/or oil storage tanks will be the responsibility of the customer.

- If any building work is required in connection with the above, a cost must be agreed prior to commencement.

7. Local authority approvals

- Any necessary local authority approvals will be sought and gained by the customer prior to commencement of the agreed work. Copies of all relevant permissions and approvals are also to be forwarded to the builder/tradesman prior to commencement of the agreed work.

- Any fees due for such approvals will paid by the customer.

8. Party Wall etc. Act 1996

- If this Act applies then it is the responsibility of the customer to follow the correct procedures. The builder/tradesman will not accept any responsibility for any delays caused.

9. Construction (Design and Management) Regulations 2007

- It is the customer's responsibility to comply with the above regulations if they apply. The builder/tradesman will not accept any responsibility for any delays caused.

10. Guarantee

- The agreed work is guaranteed against faulty workmanship and/or materials not fit for purpose for up to six years.

- An insurance-backed guarantee can be offered at an additional cost.

11. Insurance

- The customer is to notify their insurance company of the work to be carried out before commencement of the agreed work.

- The builder/tradesman will have public liability insurance of: £____________
- If relevant, the builder/tradesman will have employer's liability insurance of: £____________
- The builder/tradesman will insure the customer against theft of any materials or products taken from the premises that are for the agreed work.
- Any other cover by agreement with the customer.

Unfair Terms in Consumer Contracts Regulations 1999

The Unfair Terms in Consumer Contracts Regulations 1999 (see also page 13) apply to most terms that have not been individually negotiated in contracts with consumers. Terms that create a significant imbalance in the rights and obligations of the consumer are regarded as unfair. Unfair terms are not binding on the consumer.

Although the test in the regulations and the Unfair Contract Terms Act 1977 are not identical, it is likely that, in most cases where it is possible to apply both, the same result will occur. In other words, a term that is fair is likely to be reasonable, and vice versa.

Lead-in Time

This is the critical period between being awarded the work and commencement. It is crucial to the smooth running of the work. The length of lead-in time will depend on the complexity of the work and the current work commitments of the builder/tradesman.

For the purposes of this chapter, it will be assumed that the awarded works are a single-storey rear kitchen extension (as described in The Estimate or Quotation, see pages 28–41).

A reasonable period for the lead-in time in this case may be about six weeks. This time will need to be used wisely in checking, requesting and confirming relevant information and ordering relevant materials.

Request for written confirmation

It is advisable at this stage to request in writing the following relevant information:

- Copies of planning permission, or confirmation from the local authority that the proposed extension is a permitted development (see Local Authority Approvals, pages 59–71).
- Copies of building regulation approvals, or confirmation that the work is being carried out under a building notice scheme (see Local Authority Approvals, pages 72–75).
- Confirmation that the works do or do not need to comply with the Party Wall etc. Act 1996 (see Party Wall etc. Act 1996, pages 76–85).
- Confirmation that the work does or does not need to be carried out under the Construction (Design and Management) Regulations 2007 (see The Construction [Design and Management] Regulations 2007, pages 86–91).
- Four copies of working drawings and specification.

- For the client to inform their insurance company that they are about to start building work on their property.
- Details from the client, so that you can firm up provisional sums. (In this case, request the quotation for kitchen units so that a firm order can be placed and delivery date agreed.)
- Details from the client so that the P.C. sum can be firmed up. (In this case, request the quotation from the customer's gas supplier to move the gas meter, firm up the order with the company and agree a date for moving the gas meter. Note: Ideally this should have been done at tendering stage as often difficulty is encountered in receiving confirmation and dates when dealing with what used to be called 'statutory authorities', i.e. gas, electrical and water boards, all of which still have long lead-in times.)
- Further elaboration from the client on any unclear information provided, and on the drawings or specification, e.g. colour and type of wall and floor tiles, colour of any paint required, type of ironmongery required to doors and windows (brass, chrome or satin aluminium), electrical sockets (white, brass or chrome).
- For the client to write to their neighbours confirming that they are having building work carried out on their property and that the work will start in six weeks and will last approximately eight weeks. Ask for copies of these letters.

It is your responsibility as the homeowner to liaise thoroughly with all neighbours who might be affected by building work to your property. Do not expect your builder/tradesman to negotiate on your behalf.

Sub-contractors

If a business engages sub-contractors to carry out elements of the work, then copies of drawings and a brief specification should be sent to enable them to confirm or forward a quotation for the work. It is of course advisable to ask the sub-contractor to forward their quotation at the estimate or quotation stage (see The Estimate or Quotation, pages 27–41).

The following are elements that may be carried out by sub-contractors:

- Electricians (NICEIC registered)
- Gas work (Gas Safe registered)
- Plumbing
- Plastering
- Floor and wall tiling
- Roofing.

All sub-contractors need to confirm their estimates or quotations as well as start and completion dates. These can then be inserted into the programme for works (see Programme and Progress of the Work, pages 110–121).

Some negotiations will need to be carried out with your sub-contractor to enable your programme of work to be realistic. This is so that there are no delays waiting for sub-contractors to carry out their work, since this will not only hold up the next section of work but will also have a financial implication on the running costs of the contract.

Material scheduling

It would be advisable to prepare a schedule of all materials required for the work as soon as a contract is awarded. A percentage needs to be added on for wastage and the whole schedule should be forwarded to two or three builders' merchants, or your preferred merchant if you have a good working relationship with them, in order to establish the following:

- Best quotation for the materials (if this has not already been done at the estimating or quotation stage)
- Whether materials are readily available
- Whether any materials need to be ordered in and how long this will take

- Whether there are any specialist materials that they cannot supply
- Whether the materials are as specified or whether they are cheaper alternatives promoted by the merchant and what the potential cost savings are, if any
- Best delivery dates.

It is a good idea to check in advance what sort of plant your builder is proposing to use on your premises. For example, a mini-digger might be able to undertake the work of a full-sized digger while making a great deal less mess in your garden.

Organizing plant required for the work

Based on the site visit, it should be possible to organize the plant required. Is the plant owned or does it need to be hired, and if so at what cost? Again, it would be advisable to do this at the estimate or quotation stage (see The Estimate or Quotation, pages 27–41). It does no harm to get quotes for plant at this stage to see if there are more competitive prices available.

It would be advisable to make contact with hire companies requesting availability of plant and confirm cost.

A typical list of required plant for an average job is as follows:

- Skips or rubbish containers
- Small mechanical digger
- Scaffolding, towers and ladders
- Temporary toilet
- Acrow props and strongboys
- Breakers, hammers and grinders
- Compressor
- Lifting gear.

Whilst this list is not exclusive, it gives an indication of what plant may be required for a small contract.

Labour requirements

Now is the time to firm up your labour requirements. This is done by looking at your directly employed personnel and sub-contracted labour. The following information will need to be considered:

- Can directly employed labour be moved onto this site from another site where the contract is near to completion?
- Do you need to employ more labour and/or tradesmen? If this is the case, then time will be needed to advertise and interview.
- Can your regular sub-contractors cope with the new work?
- Do additional sub-contractors need to be sought? If this is the case, checks are required to establish that they have got the relevant tax documents and that their work is of a good standard.

Once the labour force is organized then a site foreman needs to be appointed, unless you carry out only one contract at a time and intend to be on site for most of the time.

Whatever the situation, someone will need to be on site for the following reasons:

- To liaise with the client
- To make sure the work is carried out in a safe manner (see Work on Site, page 126)
- To co-ordinate the work
- To accept delivery of materials, checking that they are correct and in good order
- To keep relevant records, including making sure that the work is on schedule (see Programme and Progress of the Work, pages 110–121)
- To ensure that the site is kept clean and tidy at all times.

Licences

Depending on the address of the site and its location, it may be that certain licences are required for any of the following:

- A licence for positioning a skip or rubbish container on the highway. Check with the local authority.
- A licence for a scaffold to be erected over a public footpath or roadway. Again, check with the local authority.

Allow enough time to contact the local authority in order to fill in relevant forms and pay any licence fee and to take note of conditions attached to the licence approval.

Ask your builder to show you all required licences for plant etc., when you are first agreeing the terms of any building work and prior to the commencement of the work.

It is advisable to set up a pre-contact meeting with the builder/tradesman to discuss basic arrangements prior to work commencing on your property. Ideally, full minutes of this meeting should be recorded and kept for subsequent reference. Points for discussion could be as follows:

- *Parking arrangements:* For site personnel as well as yourself.
- *Area for material delivery and store:* A designated area will need to be agreed and fenced off for security and health and safety reasons.
- *Position of temporary toilet:* A suitable position will need to be agreed, usually in a discreet position but accessible to the hire firm, which will include weekly emptying and cleaning within the hire costs (which are the responsibility of the builder/tradesman).
- *Access arrangements:* Vehicle movements, deliveries (health and safety issues with pedestrians walking past the premises) and your access in and out throughout the day.
- *Time of start and finish to the working day:* You should agree these times with the builder/tradesman so that they can make access available to site personnel and you will have confidence that the site will be left clean and tidy and secure at the end of the working day.
- *Lunchtime and tea breaks:* The workforce need to be aware of these times so that the entire workforce has the same lunchtime and tea break. (You can guarantee materials' deliveries will turn up just as these periods commence!)

• *Weekend working:* Is this a possibility? Will you allow this or do you want a complete break from the work over the weekend? Sometimes this can be a useful period to keep the programme on track or to allow a sub-contractor such as an electrician or plumber to keep in front of the general progress.

• *Any points of concern:* This is the opportunity for both parties to discuss openly any concerns they may have and record them in the minutes.

Conclusion

It can be seen in this chapter that the period known as the 'lead-in time' is important in order to organize the work effectively. In general, the more planning time and effort put into this period, the more successful the contract will prove to be.

The lead-in time will highlight the areas of risk and will allow the opportunity to reassess and plan for these areas.

The client will have the opportunity to discuss their concerns and will then have the time to allow you the opportunity to address them.

The contract should now get off to a good start and will hopefully proceed smoothly.

Programme and Progress of the Work

The programme for the work should form part of the contract documents. It is a document indicating the time period for an element of work. It also indicates the sequence of individual elements compared to other elements.

The programme

The programme should be easily readable and available at the site address so that all parties can see when elements of work are programmed to start and be completed.

A provisional programme can be sent to sub-contractors at the pricing stage so that they can confirm that they can achieve the dates required.

The programme can also be used to indicate any delays and the effect that the delay may have on the overall programme.

Payment schedule

The programme can be used as a schedule-for-payment chart. Each element that is completed over a four-weekly period has a value and at the end of that period a valuation of work can be assessed and agreed at the outset.

> Programming and record keeping are as important as carrying out the work itself. Not only will this keep the client informed of problems or variations, but will also give them the confidence to accept that your records are true and accurate and that payment requests are fair and reasonable. This is best practice, which will assist in keeping disputes to a minimum.

Record keeping

It must be emphasized that record keeping is one of the most important factors to consider when carrying out building contract work. In this chapter, as well as keeping a programme of work on site, several recommended forms of record sheets will be explained as follows:

Time sheet

This is to be filled in by individually employed tradesmen and labour. This sheet should show the work carried out and whether it is contract work or variation work.

Day work sheet

This sheet should be filled out by the person put in charge by the main contractor. It can be used to keep accurate records of any non-estimate/quotation work, provisional sums, P.C. sums and variations or additional work.

Allocation sheets

These are important sheets to fill in, designed to keep daily/weekly records of what went on in that week. They should indicate the following:

- Own labour on site that week
- Sub-contractors' labour on site that week
- Work carried out and percentage complete
- Information required
- Visitors to site, e.g. local authority building control surveyor, surveyors or architects, etc.
- Weather condition and hours lost due to extremely inclement weather
- Delays caused for whatever reason.

Although this chapter is primarily aimed at the builder/tradesman, it is in your interests to pay close attention to the programme and progress of building works at all times. You should also have some input on preparing the programme at an early stage.

We use the example of a single-storey kitchen extension (as indicated in The Estimate or Quotation, pages 27–41) to show the following:

- Programme of work (Table 1, see facing page and overleaf for a blank downloadable version)
- Schedule of payments showing four weekly payments (Table 2, see page 116)
- Retention held on Table 2
- Retention of 5 per cent is reduced by 50 per cent on practical completion.

If there are any delays and any variations, these can be adjusted on the above programme and payment schedule. On the following pages, examples of a typical programme and payment schedule are given. (On the tables, an 'X' denotes approximately one day's work.)

Programme of work

The programme of work allows the sub-contractors to clearly identify their work period (see facing page). It also shows the client when each element of work is to be carried out, including important sections such as breaking through from new extension, kitchen removal and insertion of new kitchen.

If there are any delays caused by weather, local authority building control instructions (additional excavations) and any variations requested by the client, these should be recorded on the weekly allocation sheets (see Table 5, page 121). It is important to keep accurate records of any of these events or instructions to show the actual effect of any of these on the programme of work.

By carrying out the above process, the client will understand the reason for delays and their effect on the programme of work. When it is explained that elements of work overrun the programme period, the client will have some understanding of the current situation.

It is always a good idea to keep fully abreast of your builder's programme of work. This is the only way to ensure everything is going smoothly and that no significant problems have arisen.

Table 1

Programme of Works for Kitchen Extension

ELEMENT OF WORK	Week 1	Week 2	Week 3	Week 4	Week 5	Week 6	Week 7	Week 8
Site set up	X X							
Demolition and removal	X			X X				
Structural alterations				X X X X				
Foundations	X X X X							
Concrete work	X	X X						
Blockwork and brickwork		X X X X X	X X X X X X					
Roofing work				X X X				
Carpentry			X X X	X X X X		X X	X X	
Plasterwork					X	X X X	X X X	X X X
Services					X X	X	X X	X
Painting						X X		X X X X X X X
Drainage	X	X X						
External work							X X	X X X X
Provisional sum for kitchen					X X X	X X	X	
Prime cost gas						X	X	
Preliminaries								

Programme of Works for Kitchen Extension

ELEMENT OF WORK	Week 1	Week 2	Week 3	Week 4	Week 5	Week 6	Week 7	Week 8

Programme and payment schedule

The payment schedule (see page 116) allows your client to plan for payments in advance, alleviating surprise when a payment request is made. It also allows you to plan their payment schedule with confidence.

If delays caused by any of the reasons mentioned above are accurately recorded, the payment schedule can be adjusted accordingly. Any variations or additional work should be recorded on a weekly time sheet (see Table 3, page 117) a day work sheet (see Table 4, page 119) or a weekly report/allocation sheet (see Table 5, page 121). These forms will provide an accurate record of additional costs or savings and will allow the client to understand any increase (or reduction) on the payment schedule at the end of each four-week period. These record sheets, if presented to the client at regular intervals, ensure that the client is not surprised by any additional costs on completion of the contract, therefore reducing the likelihood of any dispute. Examples of these three recording methods are set out below.

Weekly time sheets

It is important that each employee on site completes a weekly time sheet (see page 117) as accurately as possible. This will identify the work they do each day and the time taken, and allow you to identify any additional work carried out.

Accuracy is the key. It may be useful to offer some type of monthly bonus if time sheets are filled in accurately and on time, or for the best example of a filled-in time sheet. This will mean including this in their contract of employment.

Making regular payments to your builder/tradesman is a vital part of the transaction between you. Plan your finances in line with the programme of work so that you do not experience any surprises and your builder/tradesman does not experience any disappointments.

Table 2 **Programme and Payment Schedule for Kitchen Extension**

ELEMENT OF WORK	Week 1	Week 2	Week 3	Week 4	Week 5	Week 6	Week 7	Week 8
Site set up	X X							
Demolition and removal	X			X X				
Structural alterations				X X X X				
Foundations	X X X X							
Concrete work	X	X X						
Blockwork and brickwork		X X X X X	X X X X X X					
Roofing work				X X X				
Carpentry			X X X	X X X X		X X	X X	
Plasterwork					X	X X X	X X X	X X X
Services					X X	X	X X	X
Painting						X X		X X X X X X X
Drainage	X	X X						
External work							X X	X X X X
Provisional sum for kitchen					X X X	X X	X	
Prime cost gas						X	X	
Preliminaries								
Four-weekly payment				18,158.31				34,818.89
Less 5% retention				907.92		Less 2.5% retention		870.47
				17,250.39				33,948.42
Less previously paid								17,250.39
								16,698.03
VAT @ 17.5% (present amount)				3,018.82				2,922.16
Amount due				20,278.21				19,620.19

DOWNLOAD

Table 3

Weekly Time Sheet

Company name: ____________________

Employee's name: ____________________ Week commencing: Monday ____________________

Description of works to show time taken in hours in brackets [ ]

	06.00	13.00	13.30	16.30	Costings
Monday	Contract: ____________		Contract: ____________		
Tuesday	Contract: ____________		Contract: ____________		
Wednesday	Contract: ____________		Contract: ____________		
Thursday	Contract: ____________		Contract: ____________		
Friday	Contract: ____________		Contract: ____________		
Saturday	Contract: ____________		Contract: ____________		
Sunday	Contract: ____________		Contract: ____________		

Day work sheets

These sheets (see Table 4, opposite) are designed to record accurately time taken, materials used and plant used in an element of work that has not been included in the estimate or quotation. They can also be used to keep an accurate cost of provisional and P.C. sums. When it comes to agreeing the cost of these with the client, accurate records will be available, thus proving the actual cost.

The hours to be recorded on these sheets should be extracted from the time sheets of individual employees by the site supervisor. Copy invoices for materials and plant can be attached or reference made to invoice numbers.

The more information provided, the less objection the client can have to the cost identified.

These costs should be transferred to the payment schedule as additional works, allowing the client to identify these costs at each payment period. The client will not then be shocked or surprised on completion of the work if they are presented with an invoice for any extra work undertaken.

Table 4

Day Work Sheet

DOWNLOAD

Workman's name	M	Tu	W	Th	F	Sat	Sun	Hours	Rate	£	Expenses	Description of work	Materials, plant and transport used	Rate

Weekly report allocation sheets

No apologies are offered for repeating the key elements of this record sheet.

This sheet (see Table 5, opposite) is as important as the other record-keeping sheets because the weekly allocation sheet allows the user to look back and see what was going on during a specific week. It cannot be stressed strongly enough that accuracy is required when completing the weekly allocation sheet to identify the following:

- Own labour on site that week
- Sub-contractors' labour on site that week
- Work carried out and percentage complete
- Information required
- Visitors to site, e.g. local authority building control surveyor, surveyors and/or architects, etc.
- Weather conditions and any hours lost due to extremely inclement weather
- Delays caused for whatever reason.

The weekly allocation sheet will assist in identifying delays, cause and effect. Any delay can be transferred to an updated programme of work to identify whether a programme extension of time is required.

As soon as it becomes obvious that the programme is going to overrun (despite your best efforts to keep any delay to a minimum), a letter should be sent to the client explaining the reason, the anticipated length of the delay and the estimated new completion date.

If your builder or tradesman writes to you regarding a delay to the programme of work, ensure that you send a full reply, acknowledging the reasons given for the delay and recording your reaction to the news, together with any proposals that you might have to speed things along. This will help to indemnify you in the unfortunate event of a dispute ending up in court.

Table 5

Company Name: ______________________________ **Weekly Report**

Contract: ______________________________

Contract Duration ________ Weeks Week No.:________ Week Commencing:____________

OWN LABOUR								WORK COMPLETED	%	%	VISITORS TO SITE	
TRADE	M	T	W	T	F	S	S		Weekly	Total	Name	Represents
Total											DRAWINGS REQUIRED	
SUB-CONTRACTORS' LABOUR												
FIRM	M	T	W	T	F	S	S				MATERIALS TO BE CHASED	
											PLANT REQUIRED	
Total												

HOURS LOST AND REASON					REMARKS AND NOTES OF DELAYS
DAY	Reason	Hours	Weather	Temp.	
Monday					
Tuesday					
Wednesday					
Thursday					
Friday					
Saturday					
Sunday					
Total Hours Lost:					

Signed: ______________________________

Site Supervisor

The Work

Work on Site

In this chapter we consider the potential difficulties that both client and builder or tradesman can experience during the period of work on site. We also explore how best to set up site, and examine the health and safety issues, even if the contract is CDM 2007 exempt. (See pages 86–91 for full details regarding CDM 2007.)

As a general rule, the client is pleased to see you at the start of the job then becomes fed up with all the dust and disturbance, feeling that their home is no longer their own by midway through, and are extremely pleased to see the end of the work on completion. With this in mind, there is still the final payment to agree and receive. This chapter will explain how to reduce any potential tension between you and the client, and how to run a successful contract from start to finish.

Starting on site

The local authority building control will need to be notified in writing at least 48 hours before commencement of works. It is also a good idea to notify the planning department at the local council that the work is about to commence. In addition, the following points should be considered:

- By now the daily routine of the client should have been discussed and agreement reached on access to the premises.
- Does your client have school children? If so, make sure that your client's vehicle is not penned in by site vehicles or delivery vehicles. Your suppliers and sub-contractors need to be notified of this fact and deliveries should be arranged so as not to coincide with the school run.
- Unhindered access enabling the clients to come and go as they please will lead to a good working relationship. If the client has to chase around the site asking personnel to move vehicles, they will soon get fed up. Parking arrangements should be discussed and agreed prior to coming to the site.

- Radios are often a bone of contention between builders and clients, since they are frequently played loudly on site to the annoyance of the homeowner and their neighbours. Clients will generally accept the playing of radios so long as the volume is kept to a reasonable level. However, this issue should always be discussed and agreed in advance.
- Access to water and electricity should have been discussed and agreed prior to the commencement of the works (see The Visit, page 23).
- Protection around the site and storage areas are important. This can be provided by steel mesh fencing with concrete bases that can be hired on a section-by-section basis and erected to secure both areas.
- Has the client any pets such as cats or dogs? Arrangements should be put in place to ensure their safety.
- Temporary protection will also be required to windows and doors during the course of the work. UPVC obscure sheets are a good idea. These let plenty of light into the property but prevent the workforce from looking in on the client, accidentally or otherwise. They will also protect the glass in the windows and doors if any work is carried out close by.
- Lay temporary protection to drive and patio areas if working on or around these areas.
- If using cement mixers, select an appropriate area and provide protection from spillages.
- At the end of each working day, set aside time for cleaning and tidying up. Remember, a clean and tidy site is a safe site.

> This period during the works on site is the most testing time for the builder/tradesman and the client. Both parties naturally wish the work to be completed as soon as possible and to a reasonable standard. There are many challenges, and it is therefore necessary for both parties to act fairly, reasonably and with respect towards each other at all times.

General rules for a safe site

- Keep the site clean and tidy, with clear space for people to work in.
- Keep all people, including the public, away from danger. Fence the site off and use signs to warn passers-by.
- Ensure that structures such as walls are kept safe, and that demolition work is properly planned and workers know what to do.
- Use only 110 volt or battery-operated portable tools and safe electrical supplies.
- Make sure trenches and excavations do not collapse and ensure that people or pets cannot fall into them.
- Make sure workers cannot fall from heights; working from ladders should be allowed only as a last resort.
- Provide welfare facilities on site, dependent on agreement with the client.
- Make sure personnel on site are properly trained to do the work safely.
- High-visibility safety clothing should be worn along with hard hats, toe-protector shoes or boots, goggles when warranted and ear defenders when using noisy equipment.
- Make sure the details of the nearest hospital accident and emergency department, police and fire brigade are clearly displayed.

For further details on training or health and safety issues, contact the following:

Construction Industry Training Board (CITB) on 028 9082 5466 or at www.citbni.org.uk

Heath and Safety Executive (HSE) on 0845 345 0055 or at www.hse.gov.uk

Notifying local authority building control

We have already mentioned that the local authority building control department should be notified in writing 48 hours before the commencement of any work. Now that work is underway, the building control department will want to carry out inspections as the work proceeds.

The local authority building control department may issue cards to send in requesting an inspection. However, most authorities will accept a telephone call request. It is important not to move onto the next stage until the stage inspection has been carried out and any further work requested by the building control officer has been completed.

The stages for notification and time required are as follows:

Stage 1	Commencement (written)	2 days
Stage 2	Foundation (before concreting)	1 day
Stage 3	Foundation completed (concrete laid)	1 day
Stage 4	Damp-proof course (DPC)	1 day
Stage 5	Oversite (before concrete is laid)	1 day
Stage 6*	Before covering any structural timbers, steelwork or concrete	1 day
Stage 7	Drains in trench (before covering)	1 day
Stage 8	Drains backfilled (to be tested)	5 days
Stage 9	Occupation (if before completion)	5 days
Stage 10	Completion of work (written)	5 days

* Stage 6 is not mandatory but it is strongly recommended.

It is a legal requirement for you to notify building control and request the appropriate inspection. However, this duty is often passed on to the builder/tradesman.

While it is up to you to make sure that these inspections are carried out, it is often incorporated into the contract that the builder/tradesman should instigate these stage inspections. You should adopt a procedure to ensure these inspections have been carried out. Usually this can be done at regular site meetings, with minutes being prepared and distributed (see page 129 for advice on meetings).

Once the first inspection has been made the local authority will send you a second-stage fee to be paid. It should be noted that while the stage inspections are made, the building control officer is not supervising the work on your behalf.

If part of the existing property is being renovated then some additional inspections may be required as follows:

- Exposure of existing foundations
- Structural work exposure.

It is always worth checking with your building control officer.

Property and site security

At the end of each day, at weekends and if there are days when no personnel are on site, security needs to be carefully considered and suitable measures set in place.

- The area of the site and storage need to be kept secure whenever the site is unattended.
- If access is required to the client's premises, then arrangements need to be made to leave the premises secured at the end of each day.
- On site, the work and storage areas could be secured by using metal fencing supported on concrete bases and locked together by secure bolted brackets.
- Ladders should either be removed or stored in a safe location. Alternatively, a scaffold board laid up the rungs of the ladder and secured in place using scaffold joints and securely bolted into place will prevent access to inquisitive youngsters.
- Notices should be placed on metal fencing warning about dangers on site. Any visitors should report to the site office or the site supervisor.
- Gates should be closed on leaving site.
- The police could be notified that building work is being carried out at a given address if this is deemed necessary. They should also be advised that no work will be carried out after 6pm at night or at weekends. They may then keep an eye on the property when out on patrol. However, if any of the relevant times change, they will need to be notified.
- Valuable equipment should be removed from site or placed in a steel lock-up box bolted to a concrete base. Failing this, a good attempt should be made to hide anything of value until the building work is at a stage when it can be made secure. It is important to remember that if equipment and materials are left easily accessible and are then stolen, any insurance is invalidated.

Regular site meetings

Site meetings with the client and sub-contractors should be arranged at regular intervals. Depending on the complexity of the contract, a meeting should be arranged every two weeks or even weekly.

Accurate minutes of these meetings should be kept and distributed. Headings for these meetings should include the following criteria:

- Property address
- Date
- Those present
- Any comments from previous meetings
- Progress
- Information required by:

  Builder/tradesman

  Sub-contractor

  Client
- Any problems or concerns
- Any other business.

These minutes should be kept in a file on site so that they can be referred to at all times. The minutes should also be distributed to all relevant parties. Meetings of this kind are ideal opportunities to discuss delays and any variation in costs with the client.

It's a good idea to read the information provided in 'An overview of the main elements of a contract' (see pages 130–138) so as to gain an idea of what to expect during the contract period.

You should always look ahead at elements of work so that further instructions can be issued without delaying the builder/tradesman. Regular site meetings will assist with this.

Record keeping

In the chapter on Programme and Progress of the Work (see pages 110–121), record keeping is discussed and typical forms are indicated as follows:

- Time sheets
- Day work sheets
- Allocation sheets.

The chapter goes into some detail on the advantage of accurate record keeping and adjusting the programme of works as delays or variations in work occur.

Any additional costs should be included in the monthly valuation. On page 116 an example of a schedule of payment is shown. When the payment date approaches the additional costs can be included on an updated schedule and, because of the records kept and the meetings held, the client will be aware that there are extra costs and the records will assist with any queries.

Working to programme

The work on site should be carried out in a regular and diligent manner. This means being on site at all times (unless the weather intervenes). It also means that you will need to plan ahead and make sure that the right materials and plant are available to maintain progress as programmed. This also applies to instructed additional work. Failure to achieve this will leave you open to a deduction for damages if you fail to complete the works as programmed.

An overview of the main elements of a contract

Here, we look at the main elements of a building contract using a small single-storey rear extension as our example. This overview is not meant as a general construction guide but more of a description of each element of work and what to look out for to prevent problems arising during the construction period.

The elements detailed on pages 131–138 are taken from the programme of work as identified in the chapter on Programme and Progress of the Work (see pages 110–121).

Site set-up

This is when site huts, if they are to be used, and welfare facilities such as a temporary toilet etc., are installed in agreed positions. It may be decided to put up temporary fencing to form a material storage area as well as to fence off the site area for safety reasons and security.

The following points should also be noted:

- Protection may be required for such items as trees, garden areas, windows and doors.
- Temporary removal may be necessary of items such as television aerials and service pipes and these might also require temporary re-routing.
- Any pavers to be saved for reuse should be lifted carefully and taken to the storage area.

Demolition and removal

This element covers the demolition and removal of any outbuildings in the area of the site.

- Care should be taken to check for asbestos-bearing materials and, if in doubt, to arrange for a sample of material to be taken and tested in a laboratory. Materials to look out for are corrugated roof sheeting, cement boards such as ceiling tiles, or roof tiles and pipe insulation.
- If asbestos is found to be present, specialist contractors are required to remove and dispose of the asbestos safely. Additional costs would have to be claimed from the client. For further information on current regulations relating to asbestos, contact the HSE (refer to General rules for a safe site on page 126).
- The programme indicates removal of the outer skin of brickwork where breaking through occurs later in the programme. It also allows for removal of some inner skin bricks or blockwork to identify internal floor levels. This can often be done more easily under a window opening. If the whole external skin of brickwork is to be removed, then props and strongboys may be required and careful consideration needs to be given to the timing of this operation.

Structural alterations

This element of the work should ideally be carried out when the extension has been covered and protected from the weather.

- The work identified here is the complete removal of the two skins of external wall and the insertion of a pair of universal beams, bolted together with spacers in-between, to carry the existing external upper floor wall. The universal beams should be positioned bearing 150mm onto a precast concrete padstone at each end.
- The reveals, floor and ceiling will need to be made good and replastered and the concrete floor and screed made good (not forgetting the damp-proof membrane).

Excavations and foundations

- An initial excavation will be required to reduce ground levels. This should be to such a depth to allow for a typical floor construction as follows: 150mm of hardcore, well compacted, followed by 50mm of sand blinding, well compacted, and covered with a minimum 1200g damp-proof membrane to receive 100mm flooring grade insulation and 100mm concrete base prior to the final 75mm sand and cement floor screed.
- The total allowance for the above comes to 475mm. However, if a tiled floor finish is required, a further allowance will need to be made for the thickness of the tile and tile adhesive in order to avoid an obvious difference in height between existing floor level and new floor. Planning at this stage is very important to avoid disappointment with the final floor levels.
- Having allowed for all the above calculations, the external ground should be reduced to achieve a level floor finish and the spoil removed from site.
- Excavation for foundations needs to be at least 1m deep from ground level. The bottom of the trench should be clean and squared off and, when ready, a call made to the building control officer for a Stage 2 inspection of foundations before concreting. During the excavation stage, care should also be taken to carefully expose any mains drains, water or gas pipes and electrical cables.
- By this stage it should be known whether the works fall within a radon area. Radon is a radioactive gas that comes from natural decay of uranium found in nearly all soils. There are maps available showing where this occurs in such strength that protection needs to be incorporated into the building structure. Your local authority building control will advise further.

Concrete work

Concrete should be poured into the foundations and left approximately 450mm below ground level. This will allow for two layers of concrete blocks to be laid below ground.

- Concrete can come ready mixed and be delivered in lorries with chutes to allow the concrete to be poured directly into the foundations, or it can be mixed on site and barrowed into position. If this alternative is selected, careful batching should be undertaken to get a regular mix.
- Once poured into the foundations, the concrete should be levelled off at approximately 450mm below ground level and the building control department should be contacted to request a Stage 3 inspection.

Brickwork below DPC

The brickwork below the damp-proof course starts on the top of the foundations.

- Two skins of hollow blockwork are laid up to ground level. It is advisable to always use 7kN strength blocks at ground level, particularly if building supporting piers or if carrying onto a first-floor level (not in this example), in which the blocks can be reduced to 4kN strength.
- From ground level, the brickwork should match the existing house with either a facing brick or a blockwork and rendered finish. The damp-proof course level should be taken from the existing property. A smooth bed of mortar should be laid at the existing damp-proof course level and a new polythene damp-proof course laid and smoothed onto the mortar. Where possible, this should be laid in one continuous layer.
- On completion, the local authority building control officer should be requested to carry out a Stage 4 inspection.

Oversite

Having reduced levels, the sub-base can be laid, as described above, and the building control officer called in to carry out a Stage 5 inspection before proceeding to lay the concrete base. Having completed the oversite it is advisable to allow the concrete to go off for 48 hours before working on it.

Drains

This is often a good time to excavate and lay drain runs to include waste water and rainwater to soakaways, which should be a minimum of 5m from any building.

- Drainage could be 100mm uPVC flexible jointed pipework, which should be laid on a minimum of 100mm pea shingle and surrounded and covered with a minimum of 100mm pea shingle.
- Prior to backfilling, arrangements should be made for a water or air test to be carried out on the drain runs and for the building control officer to carry out a Stage 7 inspection. When passed, the excavations can be backfilled and connected to manholes and existing drain runs. A final test can be carried out prior to requesting a Stage 8 inspection.

Brickwork and blockwork

Prior to commencing this section, stainless steel brick vertical connectors need to be installed with bolts to the existing building.

- The external cavity wall in this element will comprise facing bricks to the external skin of the cavity wall, with a 100mm cavity to receive relevant insulation and a 100mm inner insulating block skin. Care should be taken to ensure the cavity is kept clean and free from mortar droppings. Both skins need to be tied together with stainless steel ties designed for a 100mm cavity. The ties need to be inserted at 450mm vertical centres and 900mm horizontal centres and staggered as well as being doubled up at door jambs and corners.
- The cavity at reveals of windows and doors needs to be closed by inserting an appropriate uPVC insulated cavity closure. Your local merchant will assist in the supply of this product.
- Over openings such as window and doors, steel insulated lintels need to be installed with a minimum 150mm bearing at each end. It should be noted that the lintels need to be designed to accommodate a 100mm wide cavity.
- On completion of brickwork, a 100 x 50mm treated wall plate will need to be bedded onto a sand and cement bed ready to receive the roof structure. The wall plate should be located on top of the inner skin of the blockwork and strapped down with purpose-made 1200mm long galvanized straps at 2000mm centres, screwed to the plate and inner face of the block wall.

Carpentry and joinery

The roof structure is pitched off the wall plate. In this example, a flat roof is to be constructed in the following sequence:

- Lay 150 x 50mm SC ¾ grade timbers (this grade is a structural grade and is treated) as joists for a flat roof at 400mm centres fixed to the wall plate. Joists are laid over the 3m span of the new extension and extended over the facing brick by 50mm to form a ventilation gap.
- On top of the joists, lay 75 to 19mm firring pieces to give falls and cross falls to gutters and outlets. All timbers used need to be treated.
- On top of the firring pieces, lay 19mm WBP (weather and boil proof, relating solely to the type and performance of the adhesive used to bond the layers of veneers) external quality plywood as a base for the roof finish.
- At the junction of the existing rear wall where the new roof abuts, fix a tilting fillet. This is a triangular piece of timber that assists the felt to gently turn up the wall by at least 150mm above roof level.
- At end of the joists, fix 150 x 18mm fascia board in softwood or uPVC. If timber is selected, the fascia board must be knotted and primed before fixing. There should now be a minimum 50mm air gap between the fascia and the external brick wall. A continuous uPVC ventilation strip should be inserted into the air gap.
- At the top of the fascia board, a 50 x 25mm timber drip needs to be fixed.
- The roof is now ready to be covered (see roofing section).
- The new universal beam needs studding out to receive plasterboard finish.
- Prior to covering the roof it would be wise to call in the building control officer to look at the roof structure and the universal beams in position. This would be inspection Stage 6, and whilst this is not mandatory it is strongly recommended.

Roofing work

It is now time to call in your roofing sub-contractor to prime and felt the roof.

- Is it to be a warm roof construction or a cold roof construction? A warm roof construction is when the insulation is laid on the top of the ply base and before the laying of the roof felt. A cold roof construction is where the insulation is placed in between the roof joists on the underside of the plywood, but leaving a 50mm gap between the insulation and the underside of the ply. In this case we are producing a cold roof construction.
- Felt finish can be a BS 747 felt roof in three layers finished with a capping coat, usually of a green finish. This type of felt has been used over the last 40 to 50 years. If laid properly, this material has a life span of approximately 10 to 15 years. The modern-day three-layer high-performance felts can have a life span of up to 25 years – again, if laid properly – but they are more expensive.
- The first layer should be laid directly on the ply-board base as a breather layer. No bitumen adhesive should be used. This layer is a plain layer with approximately 50mm holes to assist ventilation.
- The top layers should be dressed up the tilting fillet and bonded onto the existing brick wall. The felt should be turned up by a minimum of 150mm.
- The top layers should also be dressed down over the verge drips and dressed down the fascia timber drip by a minimum of 50mm.
- Where the felt roof abuts the brickwork of the existing house, the brick joint approximately 150mm up should be raked out and a code 4 lead flashing inserted and lead wedged in position. The lead should be dressed down the felt by a minimum of 150mm.
- The lead cover flashing should not be laid in one continuous length but in lengths of 1200mm to allow for expansion. Finally, the raked-out joints should be repointed.

Carpentry and joinery (continued)

New windows and doors can now be installed, rendering the extension and the property watertight and secure.

- Insulation to the underside of the roof is provided by 100mm insulation board, cut snugly to fit in between the joists allowing for 50mm air void between the insulation and roof board. On completion, 25mm insulation board should be fixed to the complete ceiling area before plasterboarding over
- The old kitchen units can be removed and placed in storage for clients' future use or disposed of and cleared from site. The new kitchen units and worktops (covered by a provisional sum) can be installed once plumbing and electrical alterations have been carried out.

Following on from services work as described below, allow to:

- Fit all skirting, architraves, doors and ironmongery
- Box in all pipes ready for plasterboard
- Install window sills internally
- Fit any new doors and ironmongery.

Services

Having removed the existing kitchen units, you should:

- Allow to carcass for new kitchen positions to include sink unit, washing machine, tumble drier, dishwasher and cooker.
- Allow to carcass out for new electrical sockets and lighting lay out.
- Allow repositioning of gas meter as a P.C. sum allowance; this usually entails two visits, one for the pipework and one for the gas meter.
- Following plastering and fixing of new units, allow to second fix all services, i.e. all sockets and light fittings, kitchen sink and appliances.

Plasterwork and other finishes

This is generally the messiest part of the contract. Patience is usually required on the part of the client as plaster dust settles everywhere. Further reinforcing of existing protection is required.

- Allow to plaster all walls and ceilings using Gyproc plaster backing coat and finishing coat. Allow to lay 75mm sand and cement screed to match existing surfaces.
- Allow to make good existing damaged plaster areas.
- Once the screed is dry, allow to lay wall and floor tiles to the client's selection. To assist the drying process, allow for air change by opening doors and windows and adding some heat (dependent on the time of year).

Painting and decorating

- Once dry, all newly plastered areas should be given a wash coat and two full coats of emulsion.
- New timber areas should be knotted, primed and given two undercoats and one full coat of gloss.
- Existing decorations should be made good to client's approval, unless there is an agreement for full decorations.

External work

At this point the client can generally begin to see the 'light at the end of the tunnel'.

- Make good any external grass or flowerbed areas.
- Allow to lay footpath to side and rear for access.
- Clean up site and leave clean and tidy and ready for occupation.
- Walk the completed extension and carry out a joint snagging list.
- Hand over site to client.

Delays

If any delays have occurred then these should have been recorded on the appropriate sheets and the programme of work adjusted accordingly (every effort should be made to keep delays to a minimum).

If the client is late with any required instructions and it can be proved (this should have been recorded on the record sheets) then you may have a claim for additional costs and loss and expense.

If the local authority building control officer or the client has instructed additional work, then these additional costs can be claimed within the next valuation of work and, if these are extensive and cause the contract programme to be extended, then a claim can be made for loss and expense. It must be remembered that every effort must be made to keep lost time to a minimum by attempting to make up time where possible.

Out-of-sequence working

You have taken care to programme the work carefully. If you are required to work out of sequence to achieve an element of work to suit the client and it can be proved that by carrying out this work delays have been caused or extra costs incurred, then you can make a claim for this additional amount. Again, this sequence of work should be accurately recorded on the record sheets.

It may be that one instance of out-of-sequence working will not cause too much additional time and expense but several instances of out-of-sequence working may delay the works and cause you additional costs. To prove this, accurate record keeping will be required and set against time allowed in the original programme. This additional time should be identified on the programme of work.

One method of support in proving loss is by keeping an accurate record of contract cost. This cost, if accurate, can be set against your estimated cost (see The Estimate or Quotation, pages 27–41). The estimated cost is the value without overheads and profit and the contract cost should be the same and excluding VAT. If the payments made do not cover the contract cost, then you may have a claim for loss. This will then need to be backed up by accurate record keeping showing cause and effect.

Extension of Time Issues

The contract has been signed, the go-ahead to start the works has been given and the day arrives when the works are due to commence. It is the middle of the summer and all is looking good. Work starts, and then suddenly in the first week there is a torrential rainstorm, completely stopping progress. What do you do?

After a short delay, the ground dries out sufficiently to enable you to start reducing levels and excavating foundations. Then, a redundant septic tank is exposed underground to add to your frustrations. Further delays occur while you wait for the engineer's advice and for the foundations of the project to be redesigned.

You progress onto the brickwork, but the client suddenly has a change of heart about the type of bricks originally agreed. The bricks they now want are subject to a two-week delivery lead time, causing yet further delays.

Your scaffold company was available to erect the scaffold on the day agreed on the programme, but due to all the delays they cannot now meet your revised requirements because of their existing commitments. Your job is put on hold for two or three days, causing a further delay.

So it goes on... The work continues, but is affected by delay after delay. As the work slowly progresses, it becomes obvious that you are not going to meet the programme finishing date, despite everyone's best efforts to make up time.

In your opinion the work is now delayed by about two and a half weeks.

Do try to be sympathetic if your builder/tradesman experiences difficulties that are clearly beyond their control. Sometimes the most conscientious and effective of building workers can be blown completely off course by unforeseen events. Remember, trust and good communication are essential ingredients in any successful construction project.

Extension of time issues in contracts

Under most contracts an extension of time can be claimed for the following reasons:

- *Force majeure:* This includes an Act of God and other matters outside the control of the parties, which implies force of such strength that the works could be delayed or stopped, e.g. a breakdown of electrical supplies that renders the area without electricity.

- *Exceptionally inclement weather*: It is no good claiming an extension of time because it rained for a day or two in the middle of summer. The weather needs to be extreme for the time of year and continuous. In the example given of a single-storey kitchen extension, it is suggested that the rain would need to be of storm proportions and continuous for more than a week.

The above two items are outside of your and your client's control. Therefore, an extension of time can be assessed and agreed. However, as the delay is neither parties' fault, no additional payment can be claimed. However, to claim for an extension of time, it would need to be proved that there was no alternative work to be getting on with.

As soon as it becomes apparent that the programme of works has been delayed, despite your best endeavours, confirm this in writing to your client and state how long the delay is.

If delays are caused by the builder/tradesman, then you may be entitled to deduct any cost for damages from the contract sum, particularly if an agreed damages sum was included in the contract. However, if the work was delayed by extremely inclement weather, then an extension of time can be agreed with you but no claim for additional costs and loss and expense can be made.

If the builder/tradesman makes a request for information and the required information is not covered on the drawings or specification, it should be forwarded or given to them within a reasonable period. The degree of urgency of this information should be discussed with the builder/tradesman. If they are asking for the type and colour of floor tiles then the programme will indicate when these items are required. Do not forget to allow for the delivery period when choosing tiles.

The client should be kept informed at regular site meetings of the effect of their instructions on the programme of work.

The following five circumstances can lead to a request for an extension of time, extra cost or direct loss or expense:

1 The client gives instructions to clarify their intended requirements which differ to what is shown on drawings and/or the specification of work. If this involves re-ordering materials or bringing in a further tradesman and causes delay, an extension of time can be requested.

2 The client gives instruction for additional work or for a variation in the work. Again, if this disrupts or delays the work, an extension of time can be requested.

3 An instruction from the client to postpone any elements of the work may lead to a request for an extension of time.

4 Late instructions from the client following a request by you may lead to a request for an extension of time.

5 Delays caused by late delivery of kitchen units and delays caused by the gas supplier repositioning the gas meter. Care must be taken on this subject, because if the two elements are included in the contract then it may be deemed that it is your fault for not organizing and chasing up. However, if the two elements are in the client's control, then there may be a case for claiming an extension of time.

The better the record keeping, the more chance of success the you will have for your claim for an extension of time and for additional cost, loss or expense.

As soon as it becomes reasonably apparent that the programme cannot be achieved, even taking into consideration that you have constantly used your best endeavours, then you should write to the client requesting an extension of time and indicating the extension of time required. You should also give notice in this letter of your intention to make a claim for additional work, direct loss or expense caused by the delay under the relevant sections. However, you will delay this claim until such time that relevant information is available.

For a claim for direct loss or expense to have a chance of being successful, it must be shown that this cost is not covered by any other provision in the contract and will only apply where 'regular progress' of the whole or part of the work is 'materially delayed'.

> In order for a claim to be realized, you must be able to show that you have taken all practical steps to reduce the delay.

Overleaf is a standard letter prepared as an example.

Conclusion

For the extension of time period to be taken seriously for the breakdown indicated above, accurate records will need to be kept (see Programme and Progress of the Work, pages 110–121).

If a claim for loss and expense is to have any merit, record sheets will need to be kept and accurately completed (see Programme and Progress of the Work, page 111, and Disputes and Claims, pages 164–167).

You must take into consideration that for the size of contract in the above example it is better to negotiate a reasonable settlement with the client than to resort to the courts, in which costs will soon become prohibitive.

One further advantage of keeping accurate records as far as an extension of time is concerned is that it stops the client from applying damages for late completion (see Programme and Progress of the Work, pages 110–121).

Mr Client
Address
County
Postcode

dd/mm/yyyy

Our ref: bbb/009 Ext.of.Time

Dear Mr Client,

Re: Rear single-storey kitchen extension at address.

We are pleased to advise that works are now progressing following a period of delays. However, we must also record that it has become reasonably apparent that despite this company's best endeavours, that regular progress of the work has been materially delayed for the following reasons:

1. Extremely inclement weather.

   Delay period recorded: 4 days

2. Client clarified work indicated on drawing.

   Delay period caused by re-ordering materials: 2 days

3. Additional work carried out in cleaning out septic tank, backfilling with lean mix concrete up to bottom of foundations. Construct shuttering, insert reinforcement to engineer's instruction and pour and vibrate concrete in foundations.

   Delay period caused by additional work and waiting for instructions: 6 days

4. Delays caused by late delivery of kitchen units and late installation of gas meter. Total delay period was three days but due to being under builder's control allow no delay period.

5. Delay caused by postponement of work by client for reason within his control.

   Delay period: 1 day

Therefore, extension of time request for the above is 13 working days.

Continued ...

... continued

The programme completion date should be extended by 13 working days and show a revised completion date of (insert date).

While writing we must also confirm that it is this company's intention of submitting a claim for the following:

Additional cost caused by the above items 2, 3 and 5.

Loss or expense caused by items 2, 3 and 5.

This claim will follow as soon as all the relevant information is to hand.

We will of course continue with our best endeavours to keep this and any future delays and claims to a minimum.

If we can be of any further assistance, please do not hesitate to contact us.

Yours sincerely,

Name

For company

Completion (Practical and Otherwise)

When it comes to completion, there are various different types, according to the state of the work done: completion as certified by local authority building control, practical completion and actual completion. While the completion certificate is issued by the local authority building control, the other two are different stages of completion under the terms of the contract agreed between both parties.

Completion as certified by local authority building control

This is a certificate issued to the client provided the work complies with the building regulations.

If full plans are submitted to the local authority, which are required to comply with Part 1 of the Fire Precautions (Workplace) Regulations 1997 or Part II of the same regulations, the local authority must, if satisfied, issue a completion certificate confirming compliance with fire safety requirements of the building regulations.

If on submission of a full plans application you request a completion certificate, then provided that when the works are completed all statutory requests for inspection have been carried out, the local authority building control will issue the client with a completion certificate.

Before you agree the completion of a project with your builder/tradesman, make absolutely sure that the work has been properly finished to your total satisfaction. It is a good idea to read this chapter thoroughly, as your builder/tradesman might have a very different idea from you as to when he has completed the whole job.

The local authority building control is not, however, required to issue a completion certificate if the work is carried out under a building notice procedure (see Local Authority Approvals, page 72).

> The local authority building control completion certificate is evidence (but not conclusive evidence) that the requirement of the building regulations have been complied with.

Practical completion

As explained in the chapter Negotiating the Contract (see pages 42–55) a commencement date and completion date will be agreed. Taking into consideration any extension of time agreed and allowed (see Extension of Time Issues, pages 140–145), practical completion is achieved when the extension (in this example a kitchen extension) can be handed over to the client fully operational with only minimum snagging work required. As the work is of a minor nature, it would be unrealistic to expect partial completion, particularly as the contract period is of a short timescale. It is important that the date for practical completion is recorded. This can be done by correspondence or recorded in the regular meeting minutes.

While practical completion is discussed above, in law practical completion is difficult to define and contracts do not usually qualify this point. Again, it is for both parties to act fairly and reasonably over this point.

Possession

The agreed completion date will be determined by the possession date. This is an agreed date with the client when you can start work.

The client may request that you start a day or two later for various reasons: they are not quite ready for you, or perhaps they had planned a long weekend that they had forgotten about. When you start two to three days later, you should immediately extend the agreed completion date by the same time.

> Practical completion is normally defined as when the building has reached the stage when it can be used for its intended purpose.

Consequences of practical completion

The agreement between builder and client that practical completion has been achieved is important for the following reasons:

- Retention monies held are usually reduced by 50 per cent, i.e. from the usual 5 per cent to 2.5 per cent or payment of 95 per cent upgraded to 97.5 per cent.
- Defects' liability period commences (usually three to six months).
- Builder's liability for liquidated damages ends.
- Builder's liability for certain insurances ends.

On the facing page is an example of a standard letter to the client confirming agreement of practical completion.

Completion

This is the point that you have been striving for and the client has been looking forward to. Before we reach it, and the end of the contract between both parties, we need to look at the following elements:

Defects' liability period

The defects' liability period is inserted for the benefit of both parties. It allows an agreed period of time to pass to see if any defects appear. These can then simply be corrected without any major concern. Without such a period, the client would only have redress through the courts if the builder refused to come back and rectify any faulty workmanship and/or materials.

It must be remembered that the end of the agreed defects' liability period does not necessarily mean that it is the end of your responsibility for further defects. If it can be proved that a defect has been caused by faulty workmanship and/or materials, it would still remain your responsibility.

It's good practice to always confirm in writing a verbally agreed, pratical completion date and any agreed snagging issues.

Mr Client
Address
County
Postcode

dd/mm/yyyy

Our ref: RWH/ 009 Practical Completion

Dear Mr Client,

Re: Rear kitchen extension at address.

We write to you following our meeting held on (insert date) at the above premises to confirm the following:

Practical completion was achieved (insert date).

We will now proceed to complete the list of very minor snagging issues.

Please notify your insurers that you have taken full possession of your property from the date above.

Would you please reduce the retention value by 50 per cent and forward this payment within the next 14 days.

If we can be of any further assistance, please do not hesitate to contact us.

Yours sincerely,

Name

For company

Final payment

Assuming that the defects' liability period is over and all defects have been attended to, you will now be requesting your final payment and/or release of the final retention sum. However, before this sum is paid the client should make sure that they have in their possession the following documents:

- Local authority building control completion certificate
- NICEIC electrical registration certificate for this property
- Gas Safe registration and certificate for any gas work carried out
- Warranties for materials supplied as follows:
  Windows and doors
  Kitchen units
  Kitchen appliances installed such as cooker, fridge, dishwasher, washing machine and tumble dryer
- Written guarantees from sub-contractors as follows:
  Guarantee for high-performance felt roof work.

At the handover meeting, you should hand over all of the above. It is likely at the meeting that the client will have found a defect or two still to be completed.

Arrangements should be made, in writing, to carry out these works and to confirm handover of the above documents and at the same time request final payment. The letter shown on the facing page should be sent.

Before the builder/tradesman receives his final retention payment (2.5 per cent of contract sum – refer to Programme and Progress of the Work, page 116), you should make sure of the following:

- Local authority building control completion certificate is received.
- NICEIC electrical certificate has been passed on to building control with a copy to you.
- Gas Safe registration certificate is passed on to building control with a copy to you.
- Snagging list agreed and carried out to both parties' satisfaction.
- Any manufacturers' warranties and builder's/tradesman's guarantees are handed over to you.

These items are usually organized and handed over after a short period of completion (30 days or up to three months depending on the size/complexity of the contract). On satisfactory receipt of the above, the final payment of 2.5 per cent of the contract sum can be paid.

Note: Final payment does not denote the end of the builder's responsibility. Under common law, the builder is still responsible for any faulty workmanship and/or materials for up to six years after the date of completion.

Mr Client
Address
County
Postcode

dd/mm/yyyy

Our ref: RWK/010- Completion

Dear Mr Client,

Re: Rear kitchen extension at address.

We thank you for your time in attending the handover meeting at the above property. At that meeting we presented you with the following documents:

- Local authority building control completion certificate
- NICEIC electrical registration certificate for this property
- Gas Safe registration and certificate for any gas work carried out

Warranties for materials supplied listed as follows:

- Windows and doors
- Kitchen units
- Kitchen appliances, such as cookers, fridge, dishwasher, washing machine and tumble dryer

Written guarantee from sub-contractor as follows:

- Guarantee for high-performance felt roof work

We confirm that you requested two items are attended to, which are:

- Ease and adjust internal door from kitchen to internal hallway
- Re-fix loose or hollow wall tile

Arrangements were made to complete these items on (insert day and date). Following this date completion will be achieved and this being the case we would ask that the final payment be made within 14 days of the above date. **We confirm the amount to be paid is £ (insert).**

We trust you are satisfied with the new kitchen extension you are now in possession of and that it will give you many years of useful service. If we can be of any further assistance in this or any further project, please do not hesitate to contact us.

Yours sincerely,

Name
For company

Snagging Issues

As the work comes to practical completion the builder/tradesman, client and, if engaged, a supervising officer (usually an architect or surveyor) walk round the work and produce a 'snagging list'. This is a list of items that need sorting out, adjusting or making good before final completion is achieved.

A formal snagging list should be prepared and all parties handed a copy. This should be a final list and not continuously added to, since you will need to know when the list and the work will be completed.

It must be remembered that in the construction industry tolerances are permitted and measured in +/- mm dependent on the element of work.

Shrinkage that occurs in plasterwork, carpentry and joinery is not technically a snagging item (unless it can be proved that there was excessive water content). Therefore, issues such as shrinkage cracks in plaster and timber are not your responsibility. However, so that the contract can reach completion and the client is satisfied, generally you should carry out any reasonable making good in these areas so that their final payment is made within a reasonable period of time.

Over the next few pages we will take a look at the standard of finishes for different elements of the work and give a brief checklist of what to look for. It must also be understood that while every effort is made to look at elements of the fictitious kitchen extension featured in this book, not every situation will be covered.

Snagging work is where faulty workmanship and or faulty materials have caused a problem with the finishing of the work and, rightly, should be rectified by the builder/tradesman.

Do get involved in the snagging list procedure. You know your home better than anyone and will have a heightened awareness of how things should be left upon completion of the work. Insist on including anything you are not sure about on the list; you can always talk it over with your builder later on.

External work

External walls to extension

- DPC should be at course level 150mm above ground level (this should have been inspected by the local authority building control officer).
- Any excess mortar covering facing bricks will need to be cleaned off (not with a wire brush, as this will damage the face of the facing brick).
- Any pipework coming through walls should not have a gap around it. The gap should be filled in neatly with either cut brick or matching mortar.
- Where new wall abuts existing Firfix connectors, there may be a gap or a wide joint. This could also be described as a 'movement joint'. It is advisable to fill this gap or joint with brown or appropriate colour silicone mastic. The colour of the mastic is vital here as the gap or joint should blend in with the brickwork.
- All brick joints should look the same and be pointed in the same style and mortar. Any gaps in pointing must be filled with the correct gauge mortar.
- Check for chipped or damaged bricks. These should be cut out and replaced with new bricks.
- Any efflorescence is not usually a major concern, as this will work its way out in three to five years.
- A visual inspection of the brickwork should prove that the brick courses are level and the perp (perpendicular) joints should be reasonably aligned.

Rainwater goods

- It is always useful if rainwater goods can be checked when it is raining, as any leaking joints on the guttering and downpipes will quickly become obvious.
- If the weather is fine, then inspect joints for gaps and misaligned sections.
- Check also for missing brackets and clips.
- Check that the roof verge discharges water into the gutter. If necessary, carry out a hosepipe test.

Soffit and fascia boards

- Check for gaps at abutments and at joints.
- If made in timber, check that they are already correctly decorated or that the wood has been knotted and primed ready to receive two undercoats and one full coat of gloss.

Roof

- This is a felt roof of cold roof construction. Check roof vents are in place.
- Remove any unwanted rubbish or materials from the roof.
- Check upstand, verge and flashing details.
- Make sure there are no gaps where felt layers join. There should be a minimum of 150mm overlap at the joint position.
- Test the roof's overall function with particular attention to upstand, verge and flashing details.

Pavers and remaining externals

- Check that all lawn and garden areas have been cleaned and that all rubbish and unwanted materials have been removed.
- If grassed areas are badly worn or damaged, re-seed or re-turf them.
- If fencing and gates have been used or moved and replaced, check that fencing has been replaced and properly fixed. Check that gates, latches and bolts are all working properly.
- Check paving slabs are well bedded and pointed. Make sure that there are no cracked or badly chipped pavers.

External windows and doors

- Check that you have all the keys for the windows and doors.
- Make sure that all windows and doors open and close smoothly and lock easily.
- Check that all screws have been fitted in the hinges.
- Check that when closed, doors and windows fit well and do not have parts of the door sitting proud of the frame or too close to the stop.
- Make sure externally that where the frame sits against the brickwork there is a good fit and fix against the brickwork or a good mastic seal.

- Make sure that all draught strips are correctly in position and that when doors and windows are closed there is no draught.

- Check window sills look correct and drainage details are fully functional.

- Check that trickle vents are in place and working properly.

- Check woodwork on external windows and doors is properly decorated.

Glazing

- Check that the mastic or glazing tape is properly installed.

- Check all glass areas for scratches.

- Check that all labelling is removed.

- Check that there is no paint or mastic on the glazing.

Internal inspection

Electrics

- Make sure that the work has been passed by building control or carried out by a competent electrician.

- Carry out checks on all electrical appliances.

- Carry out inspection of all sockets and lighting. Check for any damage or deep scratches.

- Check that the consumer unit has all fuses correctly in position.

Heating and plumbing

- Check that all gas work has been carried out by a Gas Safe registered engineer.

- Check that all gas appliances work.

- Receive instructions on how to operate these appliances.

- It is strongly recommended that gas detectors are fitted as recommended by building control.

- Check all sinks for any leaks. Fill sinks to check plugs hold water and release water to check waste pipes work. Place a dry finger around waste-pipe joints to check for leaks.

- Check radiators get hot when the heating is on. Sometimes air collects in radiators (this can be identified by cold spots in the radiators). Bleed radiators using an appropriate key.

- Listen for any unusual noises.
- Check pipework is insulated in internal and external areas. This will prevent heating pipes from losing excessive heat and external pipes from freezing.

Internal doors

- Check that all doors open and close efficiently.
- Check all ironmongery is in place and in good working order. Are all screws in the hinges?
- Does the door close up properly against the door stop?

Internal skirting and architraves

- Are all skirting and architraves in position? Check none are missing – particularly the short sections.
- Are all skirting and architraves lying flat and not twisted?
- Are they securely fitted?
- Are there any gaps at joints?
- Is the mastic correctly applied at the abutment of walls?

Walls and ceilings

- Check that no plasterboard nail fixings are showing through.
- Check that taped joints are not showing through.
- Check for any blemishes in plasterwork, which should be smooth.
- Check that angle bead is not visible; the solid angle bead must not protrude at corners.
- All corners should be reasonably square and upright.
- Make sure there are no plaster splashes over finishes.

Decorated surfaces

- All paint on walls and ceilings should have a solid appearance with no grinning through of undercoats.
- Make sure there are no dents or scratches in plaster surfaces.
- All woodwork should be properly knotted and primed, with two undercoats and one full coat of gloss. Appearance should be solid with a good sheen.

- Ensure there are no paint marks on finished surfaces.
- Make sure any joints have been properly filled and painted. Joints should not be noticeable.

Kitchen units

- Check that all kitchen units have been fitted properly and in the position required.
- Check that all doors open and close and do not need adjusting.
- Check that all drawers open and close smoothly.
- Check that all skirting or kick boards are in place and securely fitted.
- Check that all worktops are securely fitted and that any joints are flat and properly connected.

Wall and floor tiles

- Make sure all floor and wall tiling is flat and level.
- Make sure all joints and grouting are correctly in position.
- Grouting should run in straight lines.
- Check that the mastic filling at abutments is in place and that the correct colour has been used.
- Check that all corner beads are in place or tiles are accurately cut or abutted to form neat corners.

One thorough snagging list prepared by the client and the builder/tradesman together will, on completion of this list, bring the contract to a satisfactory conclusion.

The information provided in this chapter is a general guide to what to look out for from a snagging point of view. The list may not be exhaustive and is based on the fictitious example of a single-storey rear kitchen extension. As a general rule, a good look round will show up any snagging items required.

It is worthwhile taking a bit of time on this exercise, as one brief snagging list will be carried out by the builder/tradesman without any concern. However, if the list is continually added to and extended, the builder/tradesman will soon get fed up as he or she will need to keep going over work that they have previously carried out. This could even cause out-of-sequence working, which is to no one's advantage (for more information, see page 139).

Disputes and Claims

If there is a written agreement between both parties, variations to the work have been recorded in writing and both parties are communicating well, all as described within this book, then the risk of a dispute and subsequent claim for loss and expense will have been significantly reduced.

Unfortunately, because of the difficulties that can be encountered within the building environment, there is always the possibility of a dispute arising.

If a building contract has been used it usually has a mediation, conciliation or an arbitration clause. If not, your dispute could end up in the county court by way of the small claims track, fast track or multi track. The track will depend on the value and complexity of the claim.

Reasons for disputes

Due to the complexity of issues, and particularly in the current economic climate, disputes can arise and may be about the following:

- Quality of work
- Time taken
- Price (particularly extra costs)
- Quality of materials.

Remember, using the court process should be the last resort. There are numerous ways to solve disputes and these are called Alternative Dispute Resolution (ADR) schemes (see pages 161–163).

Under the Supply of Goods and Services Act 1982 you are entitled to expect the following standards from the builder or tradesman:

Quality of work

All services must be carried out with reasonable care and skill. If a main contractor is employed along with a sub-contractor, the main contractor is responsible for the quality of the sub-contractor's work.

Time taken

If a tradesman did not set a clear date to start and complete the works, the above Act expects the work to be finished within a reasonable time. If, however, a completion date was agreed and the tradesmen fail to achieve this (without reason for an extension of time) the contract has been broken and you are entitled to compensation for any loss you may have suffered.

Price

It is usual to agree a price at the outset. Sometimes this is not always possible. The above Act says that no more than a reasonable charge should be made for the work. You can assess what a reasonable charge is by asking other builders/tradesmen. If a firm price or a fixed hourly rate has been agreed and there has been no variation in the work, then both parties are bound by it.

Materials

Any materials supplied as part of the works must be described by the builder/tradesman, and be of satisfactory quality and fit for the purpose for which they are intended. If they are not, you are entitled to compensation. Be aware that if you supplied the materials and insisted on using them, then you may not be able to claim any compensation from the tradesmen.

To summarize, the Supply of Goods and Services Act 1982 requires a supplier of services acting in the course of his business to carry out that service with reasonable care and skill and, unless agreed to the contrary, within a reasonable timeframe. He should also make no more than a reasonable charge.

Over the following paragraphs the ADR schemes and county-court track system will be explained in detail as follows.

Dispute process

If a dispute arises between you and the homeowner, and it cannot be settled, then the following options are available to you:

- The legal process via the courts
- Arbitration
- Adjudication
- Conciliation and mediation.

The legal process via the courts

It can be a lengthy and costly route to gain a county-court judgement. This process is open to the public and may be reported on in the press or by other means.

County-court rules require that ADR is considered before the final step of using the courts is taken.

The route the case follows is decided by the judge and is based on the value of the claim and how complex the case is. It affects everything from how the case should be prepared, to the length of the hearing and even the type of judge.

There are three routes to take, which are called tracks, as follows:

Small claims track: generally for lower value and less complex claims with a value of up to £5,000.

Fast track: for claims with a value of between £5,000 and £25,000.

Multi track: for more complex claims with a value of £25,000 or more.

Alternative Dispute Resolution (ADR)

Alternative Dispute Resolution is any type of procedure or combination of procedures voluntarily used to resolve disputes.

The main types of ADR used in building disputes are as described overleaf. Mediation and conciliation are often provided by trade associations.

The parties who decide to use ADR to settle their disputes can select a method and a provider of their own choice. As mentioned previously, the civil-procedure rules provide for the judiciary to encourage the use of ADR in appropriate cases.

Any particular terms required, such as confidentiality clauses, should be agreed at the outset.

Some of the advantages of using ADR are as follows:

- A change in the way a person or organization behaves
- An agreement to replace or repair
- A mistake corrected or compensation
- An apology
- An explanation of what happened
- Length of time to resolve dispute.

Courts may offer the following:

- An order that something be done or stopped
- Compensation
- A judgement from the court about who is right and who is wrong
- Enforcement procedures of any judgement.

Think carefully before using ADR and consider the following:

- The outcome you want (result)
- What you can expect to receive
- How you want to go about solving your problem
- How willing the other side is to try to resolve the problem.

Arbitration
This process involves an independent adjudicator or arbitrator (someone who does not take sides, and who will not gain or lose anything by the outcome). He or she hears both sides of the disagreement and may receive relevant paperwork, then makes a decision that will solve the problem. In most cases the arbitrator's decision is legally binding on both sides.

Arbitration, if both parties agree, does not need to involve hearings and can be decided on documents alone.

Both parties must agree to use this process.

This process is confidential and so is the amount of compensation that is awarded.

> Arbitration is less formal than a court hearing but in many ways is similar, with procedural rules that govern issues such as disclosure of documents and evidence. The process is private rather than public.

Adjudication
The Housing Grants Construction and Regeneration Act, Part II Scheme for Construction Contracts provides for adjudication under any contract in writing, irrespective of whether that contract may include an adjudication provision. In short, it is the law. However, the Act does not extend to:

- Contracts that are not in writing
- Contracts made with a residential occupier
- PFI contracts.

Nevertheless, if the contract/agreement includes an adjudication clause or is covered by the 'scheme', then that would be the first port of call in any dispute. In this event, a court of law would not normally deal with the dispute and refer the parties to adjudication. An adjudication decision is normally binding on the parties.

Adjudication involves an independent and impartial third party, who considers the claims of both sides and makes a decision. The adjudicator is usually an expert in the field. Adjudicators are not bound by rules of litigation and arbitration.

Within the Housing Grants and Regeneration Act 1996 most building contracts should include an adjudication clause.

> Adjudication is less formal than arbitration. Adjudication is not always binding on the parties, so the case can still be referred to the courts if required. Arbitration, on the other hand, is binding on both parties and can only be referred to the courts if there is an obvious legal mistake or the arbitrator has behaved improperly.

Mediation and conciliation

Mediation and conciliation are similar. This process involves an independent person to listen to the issues. Mediation can be heard by any independent person, whereas conciliation should be heard by an experienced person in the field of the dispute.

Mediators or conciliators will help both parties find a solution to their problem. Both parties involved in the dispute decide on the outcome of the dispute and any terms of agreement (not the mediator or conciliator).

> In the mediation and conciliation process it is up to both parties to make the decisions guided by an independent person. Both parties must agree to this process. This process is confidential, so cannot be used in court later unless both parties agree.

Claims

Depending on the nature of the dispute, you could consider making a claim for loss of profit and loss and expense; the Head of Claim may be as follows:

- Loss of profit on work carried out (not indicated in the claim below)
- Claim for loss of profit on work still to be carried out
- Claim for delay
- Claim for disruption.

For any claim to have a chance of success, cause and effect must be shown. This is the reason for keeping strict records and showing the effect the client's changes may have on the contract programme (see Programme and Progress of the Work, pages 110–121).

Reasons for a claim may be as follows:

- Client making changes as work proceeds and delays and disrupts the sequence of work
- Out-of-sequence working
- Due to late instructions you were unable to work regularly
- Additional works
- Refusal to let you back on premises to complete and rectify any snagging issues.

Any of the above could have occurred because the client kept changing their mind on the work and continually disrupted the work with amended instructions and/or issuing late instructions for additional work.

This could have the effect of a delayed finish and a request for additional payments. If this effect was not fully explained to the client and proved by the use of the previously described record sheets (see pages 112 and 121) and the effect identified on the programme of work, then the client might consider that the works are progressing too slowly (two and a half weeks behind programme). In these circumstances, the client may consider your request for additional costs inflammatory, leading to a possible dispute and you being asked to leave the site with the work 70 per cent complete.

A claim scenario based on the above Head of Claim could be as below and overleaf.

Claim 1. Loss/expense caused by disruption

Value of work carried out based on 70 per cent of work completed.

Contract sum = £34,870.84 x 70 per cent = £24,409.59.

Cost of carrying out works to date from contract accounts = £27,880.55.

Loss on contract to date = £ 3,470.96.

While this is a start, to show loss in itself is not the only evidence necessary to prove that the dispute was fully caused by the client.

So the record sheets should show elemental work and items or phrases such as:

1. Stop work on brickwork, waiting for client's instruction.
2. Out of sequence on carpentry work due to client's late instruction.
3. Altering work due to client's change of mind.

Time recorded against items 1, 2 and 3 = 47 hours for two labourers =
Total of 94 hours at £15.00 = £1,410.00

Time recorded against items 1, 2 and 3 = 39 hours for three tradesmen =
Total of 117 hours at £18.50 = £2,164.50.

Work out estimated cost for carrying out these works in your elemental breakdown of say £2,245.50

Actual cost of work caused by disruption = £1,410.00 + £2,164.50	= £3,574.50
Cost of work in elemental breakdown of estimate/quotation	= £2,245.50
Loss caused by disruption of labour	**= £1,329.00**

Claim 2. Delay

Additional preliminary (taken from The Estimate or Quotation, see pages 29–30) cost as follows:

Supervision = £444.00 divided by 8 x 2.5 weeks	= £ 138.75
Scaffold = £135.00 per month, therefore divide by 4 x 2.5 weeks	= £ 84.38
Insurance = £950.00 cost per year, divide by 52 = £18.27 x 2.5 weeks	= £ 45.68
Temporary WC weekly cost = £30.00 therefore x by 2.5 weeks= £ 75.00	
Phone = £16.59 per week x 2.5 weeks	= £ 41.48
Additional cost of preliminaries caused by delay	**= £ 385.29**

Additional cost of labour and plant caused by delay:

Bricklayer 3 days at £150.00	= £ 450.00
Labourer 3 days at £120.00	= £ 360.00
Mixer 3 days at £22.50	= £ 67.50
Additional cost of labour and plant caused by delay	**= £ 877.50**
Total cost of claim caused by delay	**= £ 1,262.79**

Claim 3. Loss of profit on work not carried out

Contract sum	= £34,870.84
Value of work when asked to leave site	= £24,409.59
Work outstanding	= £10,461.25
Therefore loss of profit on £10,461.25 value of work. Taking into consideration that 20 per cent profit is built in, therefore cost	= £ 8,717.71
Profit is £10,461.25 less cost at £8,717.71 = profit of	= £ 1,743.50
Loss of profit on work not carried out	**= £ 1,743.50**

Summary of claim

Claim 1 Loss/expense caused by disruption	= £ 1,329.00
Claim 2 Loss caused by delay	= £ 1,262.79
Claim 3 Loss of profit on work not carried out	= £ 1,743.50
Total claim for loss and expense and profit	**= £ 4,335.29**

Conclusion

As can be seen from the preceding explanation, it is vitally important to keep accurate records to prove your claim for loss/expense and profit.

A claim should be accurately produced from the record sheets and as indicated on the programme and any revised programme. Do not forget to confirm instructions and the reasons for delays in writing as soon as they become basically clear.

In the example provided here it is considered that the client is in breach of contract by preventing you from completing the work on site. It is important to prove this breach by correspondence and stating that you are ready, willing and able to return to site and complete the work.

It is also important that the client has a full explanation as to why and how the programmed work has slipped (see Programme and Progress of the Work, pages 110–121).

Proving breach of contract can be difficult, but keeping accurate records and minutes will help.

The best advice is to try to negotiate a settlement figure with the client. Using the courts can be a long and expensive process and no matter how much it is believed that a good case is in hand, the courts and the cost of their service can always be a surprise. Consequently, both you and the client should think carefully before resorting to the courts.

The example given here is by no means exclusive but hopefully gives you an understanding of a claim and how it is made.

Appendices

Housing Grants, Construction and Regeneration Act 1996

The Housing Grants, Construction and Regeneration Act 1996 is an Act to make provision for grants and other assistance for housing purposes and about action in relation to unfit housing to amend the law relating to construction contracts and architects; to provide grants and other assistance for regeneration and development and in connection with clearance areas; to amend the provisions relating to home energy efficiency schemes; to make provision in connection with the dissolution of urban development corporations, housing action trusts and the Commission for the New Towns; and for connected purposes.

The Act is formed in five parts as follows:

Part I: Grants etc. for renewal of private sector housing

Part II: Construction contracts

Part III: Architects

Part IV: Grants etc. for regeneration, development and relocation

Part V: Miscellaneous and general provisions.

In this addendum we examine Part II, Construction contracts.

The Housing Grants, Construction and Regeneration Act 1996 introduced a series of rules for qualifying contracts entered into after May 1998. Section 107 of the Act sets out which contracts will qualify as 'construction contracts'. To explain:

It must be an agreement either in writing, or evidenced by writing, for the carrying out of construction operations. Certain contracts such as the delivery of materials to site and contracts for residential occupiers are excluded.

Parties to a contract are free to negotiate their own terms and conditions under which the work is to be carried out. There are times, however, when the contract fails to comply with the minimum requirements relating to adjudication and payment laid down by the Act. Consequently certain provisions will be automatically incorporated into the contract by the scheme for construction contracts.

Summary of minimum requirements

When does the Act apply?

- Contracts in writing or evidenced in writing.
- Contracts for construction operations such as alterations, repairs, maintenance, decorations, demolition and installation of buildings forming or to form part of the land.

Adjudication

Incorporated within the Act is the Scheme for Construction Contracts (England & Wales) Regulations 1998, otherwise referred to as 'The Scheme'. This sets out the rules and procedure for adjudication.

Irrespective of whether any contract includes a clause or any other reference to adjudication, all contracts must:

- Allow either party to refer a dispute to adjudication.
- Provide for an adjudicator to be appointed within seven days.
- Require the adjudicator to reach their decision within 28 days. Longer periods may be allowed if agreed by both parties.
- Allow the adjudicator and the party who referred the dispute to extend the period for their decision by up to 14 days.

- Impose a duty on the adjudicator to act impartially.
- Enable the adjudicator to take the initiative in ascertaining the facts and the law surrounding the dispute.
- Provide for the decision of the adjudicator to be binding on the parties or until the dispute is taken to further adjudication and/or the law surrounding this dispute.
- Provide that the adjudicator is not liable for anything they do unless they act in bad faith.

If the contract does not comply with all eight elements in full, all the adjudication provisions of the contract will be set aside. The provision of the Act will apply and set time limits on all eight points above.

Payment

The Act and the scheme for construction contracts provide a number of payment provisions covering:

- Payment by instalments
- Final payment
- Withholding payment
- Conditional payment.

If the contract fails to comply with any one of the above provisions, the provision from the scheme will apply.

Suspending performances

If any payment is not received by the final date for payment and a notice of withholding payment has not been served, the contractor may suspend work after giving seven days' notice. The right to suspend performance is ceased once payment has been received. The period to complete the work is automatically extended by the required period of suspension.

> This brief summary of the Housing Grants, Construction and Regeneration Act 1996 is not intended to be a detailed explanation of the provision of the Act. It would be wise to seek legal advice on any specific issues.

Sustainable Construction

The United Kingdom government, having signed up to the Kyoto Protocol, is putting pressure on all parties in the construction industry to reach binding targets on reducing carbon emissions. Thirty-seven industrial countries and the European Community have targets for reducing greenhouse-gas emissions.

The government is actively encouraging and promoting sustainable construction in order to reduce the United Kingdom's carbon-emission output from housing and construction. The construction industry will need to adapt and change to meet new requirements and regulations.

As sustainability is still a relatively new concept, developers and builders are lacking the knowledge to fully understand sustainable construction.

Due to pressure from the government through the Code for Sustainable Homes, 2007 (CSH) and the gradual increase and improvement in building regulations, the public, developers and builders will soon become more aware of sustainable construction.

A generally accepted definition of the concept of sustainability is to ensure that development seeks to meet the needs and aspirations of the present without compromising the needs of tomorrow. For the construction industry, this means delivering buildings and communities with lower environmental impact while enhancing health, productivity, community and quality of life.

The Code for Sustainable Homes, 2007 has been introduced by the government to give ratings for the amount of carbon emission that the development produces. The government has set a target to achieve zero-carbon status for new-build housing by 2016 and all buildings by 2019.

New legislation regulations and guidelines will evolve and the construction industry will need to evolve with them if it is to meet the new government targets. The government has implemented an impact assessment on relevant building regulations. This demonstrates the government's intention to improve and apply new building regulations in order to meet strict targets.

The Code for Sustainable Homes, 2007

The Code for Sustainable Homes has a six-star rating that will assess each house over the following nine key issues:

1. Water efficiency
2. Surface water management
3. Site waste management
4. Energy efficiency
5. Pollution
6. Health and wellbeing
7. Management
8. Ecology
9. Materials.

Speak to your local builders' merchants to find out which materials comply with The Code for Sustainable Homes, 2007.

The Code for Sustainable Homes, 2007 is assessed in two phases:

Phase 1

An initial assessment and interim certification is carried out at the design stage. This is based on design drawings, specification and commitments that result in an interim certificate of compliance.

Phase 2

A final assessment and certification is carried out on completion of the construction work. This is based on the design-stage review and includes a confirmation of compliance, with site records and visual inspection.

Building regulations

Over the next few years building regulations will be adapted and improved to assist the government in reaching its targets.

Recently there has been a consultation on the 2010 version of Part L: this part deals with energy use. The previous two revisions of Part L in 2002 and 2006 were the cause of many problems, in particular understanding the Standard Assessment Procedure (SAP) calculations. Consequently, it is the government's intention to avoid as many problems as possible in the 2010 version.

Part F is linked to Part L so the consultation will cover Part F as well.

Part G is called 'Sanitation and Water Efficiency' and is due to be published imminently and may take effect from October 2010. This was due out in April 2010 but was delayed due to changes.

Other parts of the building regulations will be updated at fixed intervals. There will be changes to Part A, which deals with structures, Part C, which covers site preparation and moisture, and Part J, which regulates heat producing appliances. Consultation papers will need to be published on these parts.

It will not be too long before these regulations, in their new form, will be part of new building projects.

Other methods used to assist the government in reaching its targets

Planning Policy Statement 22 (PPS22) Renewable Energy

This enables local authorities to demand that a percentage of a building's energy requirement be generated using on-site renewable energy (such as solar panels to generate electricity in order to provide hot water, or biomass pellet boilers to provide heating) as a condition of planning permission. The percentage is variable and, typically, local authorities may demand between 10 and 20 per cent.

Energy certification of buildings

All new and existing buildings need to be assessed for energy efficiency before being sold or rented. The building fabric needs to be assessed by an accredited inspector and given an energy rating on an A to G scale with A being the best. Buildings over 1,000m^2 occupied by public bodies or visited by a large number of the public need to display a certificate showing how much energy is being used. Compulsory efficiency checks for air-conditioning are also being introduced.

Site waste management plans

Site waste management plans are compulsory for any construction project with an estimated cost of £300,000, with slightly different rules for works costing more than £500,000. The idea is that the process will make contractors aware of how much waste they are producing, prompting them to reduce it. One person is made responsible for estimating and recording waste produced and recycled during the job. The plan must be regularly updated.

Glossary

ABI Association of British Insurers

Abutment Structure that supports the lateral pressure of an arch or span

Adjudication Alternative Dispute Resolution involving an independent and impartial third party who considers the claims of both sides then makes a decision. It is less formal than arbitration. Ajudication is not always binding on the parties, so the case can still be referred to the courts if required

Allocation sheet Record of the works carried out on site in a specific week

Alternative Dispute Resolution (ADR) Procedure or combination of procedures voluntarily used to resolve disputes. *See* Arbitration, Concilliation and Mediation

Angle bead Metal or uPVC section used to create corners

Arbitration Alternative Dispute Resolution involving an independent ajudicator or arbitrator who makes a decision that is, in most cases, legally binding on both sides. It is less formal than a court hearing but has similar procedural rules

Architectural salvage yard Yard where second-hand materials can be purchased

Architraves Moulded timber used to cover joints between door lining/frame and plasterwork

Area of Outstanding Natural Beauty (AONB) Area designated by the local authority as having significant landscape value

BSI British Standards Institute, i.e. the National Standards Body in the United Kingdom

Bills of quantities Lists of quantities required for building work

Biomass pellet boilers Eco-friendly timber pellets used as fuel for boilers

Bitumen Black, sticky, tar-like form of petroleum

Boxing in Cover for service pipes and cables

Building Notice Alternative method of L.A.A. Building Regulation Compliance, which does not require the submission of detailed plans. It is suitable for small, uncomplicated works

Building Regulations approval Compliance with local authority regulations

CDM Coordinator Role to advise client on health and safety issues

CDM Regulations 2007 Construction (Design and Management) Regulations 2007

CERTASS Government-approved Quality and Evaluation Scheme for Certification and self-assessment

CSH Code for Sustainable Homes

Casement Part of window

Cavity construction Two skins of wall

Cavity walling External wall of building in two skins with cavity in between

Cladding Finish to the face of a building

Cold roof construction Roof insulated within the roof void

Competent Person Scheme Government scheme allowing individuals or enterprises to self-certify that their work complies with Building Regulations

Completion Certificate Certificate issued on completion by Local Authority Building Control and/or a Supervising Officer

Completion (Practical) Programme of work that is all but complete except for minor snagging work to do

Conciliation Alternative Dispute Resolution involving an independent third party that is experienced in the field of the dispute. A conciliator will guide both parties to reach an agreement

Construction Industry Training Board (CITB) Organisation that assists with training, including health and safety

Contract An agreement between two parties

Core term Term defining product and price

Curtilage An area of land surrounding a building

Damp proof course (DPC) Used to prevent moisture from the ground rising into the internal fabric of a structure

Daywork sheet Record sheet to note down elements of work that are not fully priced for

Defects liability period Agreed period of time after practical completion to address defects

Design and Access Statement Statement on design and access for building in Green Belt, Conservation or listed buildings

Designated land Includes National Parks, the Broads, areas of Outstanding Natural Beauty.

Downpipe Vertical pipe taking rainwater from gutters

Duty holder A person with legal duties under CDM Regulations 2007

Eaves height Lowest point of pitched roof

Efflorescence Type of salts appearing on face of brickwork

Elemental estimate or quotation Breakdown of costs into items or trades

Employers Liability Insurance Type of insurance covering employees for compensation (compulsory)

Energy certificate Assessment of energy rating by accredited inspectors

Escape window Window with an opening casement large enough to allow for means of escape

Estimate An approximate price given based on minimal information that is not binding on either party

Extension of time Programme of work extended for good reason

FENSA The standard for replacement windows

Fanlight Small opener usually at the top of a window

Fascia board Timber fixed to end of joists to carry gutter

Fast track Procedure for claims of between £5000 and £25000

Firring piece Timber cut to give falls

Flashing Strip of metal, usually lead, used to stop water penetrating the junction of a roof with another surface, such as the chimney

Formal contract or agreement Written agreement or contract between parties

Full planning application An application to the Local Authority for permission to build or for change of use

Gas Safe Registration Competent person in Gas appliances

Grinning through Where an earlier coat of paint can be seen through a top coat

Guidance notes Advice on completing an agreement

HSE Health and Safety Executive

Hardcore Broken brick or rubble under concrete base

Hardstanding Concrete base to stand on or for a vehicle to be parked

Head of Claim Heading for each claim made

Ironmongery Door or window handle, hinges and lock

Joinery Finished timber, such as stairs, cabinets and wardrobes

Kickboards Turned-up scaffold board at edge of platform on scaffolding

Knotting Product painted onto knots in wood to stop resin bleeding

Lath and plaster An early method of plastering walls incorporating timber slats and lime plaster

Lean mix A weak mix of concrete

Lintel Steel or reinforced beam over opening

Listed Building Consent Building listed Grade 1 or Grade 2, of particular note for period or age

MDF Medium density fibreboard

Means of escape Route of escape in case of fire

Mediation Alternative Dispute Resolution similar to Conciliation but differing in that the mediator can be any independent person

Moisture ingress Where water finds a path into a building

Movement joint A flexible joint used to reduce the possibility of cracking

Multi track Procedure for claims of more than £25000

NICEIC National Inspection Council for Electrical Installation Contractors

Notification If required, notification of work to the Health and Safety Executive

Outline Planning Permission (OPP) Initial view that will need a further application

Out-of-sequence working When work is carried out in the wrong order

Overheads The hidden costs of running a business

Oversite Area of ground floor slab – preparation and sub-base

P.C. sum Prime cost sum for works or services provided by nominated subcontractor

P.P.S 22 Planning Policy Statement 22

Padstone Concrete block to support each end of a steel beam

Party fence wall Wall on a boundary line but not part of the building

Party Wall etc. Act 1996 An Act of Parliament to protect your neighbour's property and your right to carry out work

Permitted development Development that falls within a set criteria in the Planning Act

Planning breach Failure to obtain planning permission

Planning permission Local Authority Approval

Plasterboard A papered panel containing gypsum that is fixed to a wall or ceiling

Possession date A date given to take possession of a building

Preliminaries Items included in a contract sum that cannot be satisfactorily split between elemental costs

Profit A sum made over and above the cost

Protected species Species of animal or bird that is protected by law

Provisional sum A sum allowed for work that cannot be entirely foreseen

Public Indemnity Insurance Insurance against personal injury claim or property damage

Quotation A fixed sum for work or a product

RSJ Rolled steel joist used to support openings

Radon A natural gas occurring during decay (radioactive)

Rate Cost per cubic, square or linear metre

Retention sum Sum held by client to cover defects

Reveal Short width of wall up to window or door

Ridge height Usually the highest point of a roof

Safety glazing Glass that will not splinter if broken

Schedule of Conditions Indicates responsibilities and further explanations

Screed A sand and cement finish to floor

Second fix Fix of final element of work

Serving Notice Giving notice to a neighbour about intended work

Silicone mastic A flexible filler at abutments

Skirting Finished timber section at junction of wall and floor

Small claim track Claim for under £5000

Snagging list List of minor works or repairs prior to completion

Soffit boards UPVC or timber under fascia board

Statutory Authorities Gas, electricity, water boards

Stopcocks Mechanical gates to stop or allow the flow of water or gas

Strong boys Steel plates to support loads

Studding out Timber section to form walls or boxing

Subcontractor Company engaged on element of work by Main Contractor

Surveyor's Award Party wall award including records of condition

Sustainable construction Environmentally responsible building

Terms and conditions Terms and conditions under which a company or person works

Tilting fillet Timber, often triangular, that is used at the eave and gable to begin the roof at the correct angle

Time sheets Sheets filled in by worker to indicate time taken to complete work

Tree Preservation Order Form of planning control for trees

Trickle ventilation Small slots in windows or walls that can be reduced or closed to allow ventilation

Universal beam Steel beam to support openings

U-value A measure of how well a building component keeps heat inside a building

Variation Where work is varied leading to an adjustment in cost

WBP Weather and boil proof, relating to glue bonding plywood

Wall ties Ties both skins of cavity wall

Warm roof construction Insulation on top of roof timbers

Warranties Insurance-backed guarantees

Welfare facilities Canteen and toilet facilities for workforce

Resources

Useful websites

www.businesslink.gov.uk (Practical advice for businesses – the official government website)

www.bis.gov.uk (Department for Business Innovation and skills)

www.cskills.org (Sector Skills Council and Industry Training Board for the construction industry)

www.fsa.gov.uk (The Financial Services Authority)

www.hse.gov.uk (Health and Safety Executive)

www.oft.gov.uk (Office of Fair Trading)

www.opsi.gov.uk (Office of Public Sector Information)

www.planningportal.gov.uk (Online planning and building resource)

Explanatory booklets

Building Regulations, Office of the Deputy Prime Minister

Party Wall etc. Act 1996, Department of Environment Transport and The Regions

Acts and regulations referred to

Housing Grants, Construction and Regeneration Act 1996

Party Wall etc. Act 1996

Sale of Goods Act 1979

Sustainable and Secure Buildings Act 2004

The Building Act 1984

The Town and Country Planning Act 1990 (as amended by the Planning and Compulsory Purchase Act 2004)

Unfair Terms in Consumer Contracts Regulations 1999

Index